MARINE BIOLOGY

PORPHYRA YEZOENSIS

FRONTIERS IN PHYSIOLOGICAL AND MOLECULAR BIOLOGICAL RESEARCH

MARINE BIOLOGY

Additional books in this series can be found on Nova's website under the Series tab.

Additional E-books in this series can be found on Nova's website under the E-books tab.

MARINE BIOLOGY

PORPHYRA YEZOENSIS

FRONTIERS IN PHYSIOLOGICAL AND MOLECULAR BIOLOGICAL RESEARCH

KOJI MIKAMI
EDITOR

Nova Science Publishers, Inc.
New York

LIBRARY OF CONGRESS CATALOGING-IN-PUBLICATION DATA

Mikami, Koji.
Porphyra yezoensis : frontiers in physiological and molecular biological research / author, Koji Mikami.
p. cm.
Includes index.
ISBN 978-1-61122-304-0 (hardcover)
1. Porphyra--Physiology. 2. Porphyra--Molecular aspects. 3. Porphyra--Research. I. Title.
QK569.B2.M55 2010
579.8'9--dc22

2010038650

Published by Nova Science Publishers, Inc. † New York

CONTENTS

PREFACE

Red algae diverged into unicellular and multicellular lineages and dispersed throughout marine environments and, to a lesser extent, to fresh water and extreme environments. *Porphyra* species are red seaweed taxonomically belonging to the Bangiophycideae of Rhodophyta and are important marine crop species. They have been utilized as food and medical resources since ancient times in several Asian countries, especially Japan, Korea and China. In Japan, *Porphyra yezoensis* was identified by Professor Kichisaburo Endo (Hokkaido University) in 1908, in Hakodate City (Hokkaido, Japan). Since then, *P. yezoensis* has become the representative culture species, and it has been extensively bred to produce the high-quality product Nori. Due to its economic importance in Japan, basic and applied biological studies of *P. yezoensis* have been carried out primarily by Japanese scientists.

Bangiophycean algae, including *Porphyra* species, are unique and important in the field of experimental biology. For instance, the fossil record of an ancient Bangiophycean algae, *Bangiomorpha pubescens*, is dated at 1,200 MYA and shows multicellularity, cell differentiation, and both sexual and asexual life cycles. Thus, Bangiophycean algae are an attractive experimental model for studying the origin and regulation of life cycles of multicellular organisms and sexual differentiation in eukaryotes. In spite of this, cellular and molecular biological approaches have not yet been extensively applied for red algal studies, due to the lack of modern experimental systems routinely used in other eukaryotes. However, it is worth noting that the recent establishment of a standard laboratory culture strain enables the preparation of large quantities of isogenetic material for biochemical analysis and nucleotide extraction. It also enables the development of transient transformation systems, an important technical breakthrough for modern seaweed research. Moreover, large-scale EST collections and whole genome sequencing projects are now underway. These technical improvements and the accumulation of genome-wide information have stimulated cellular and molecular biological studies of *P. yezoensis*. The recent advances in the field could provide useful genetic resources to elucidate the life cycle and environmental responses of multicellular red algae and may also be used for biotechnological applications, such as molecular breeding and the discovery of novel functional materials.

The aim of this book is to present an overview of current research activities in *P. yezoensis* by Japanese scientists. Reviews of the latest research in the growing fields of study are gathered with a focus on topics such as the improvement of experimental systems, the physiology and molecular biology of development and stress responses, new approaches for breeding, the use of genes in biotechnology, and characterization of functional materials. The editor hopes readers will enjoy the excellent reviews provided by authors who have kindly contributed their time for this book. In addition, the editor hopes that this book stimulates progress in the molecular biological research of *P. yezoensis*, as well as the development of this organism as a global model species in biotechnology, through the sharing of information necessary for the further progress of red algal research.

Koji Mikami
Hakodate, 2010

In: Porphyra yezoensis
Editor: Koji Mikami

ISBN 978-1-61122-304-0

Chapter 1

PORPHYRA: MODEL PLANTS IN MARINE SCIENCES

Naotsune Saga*

Faculty of Fisheries Sciences, Hokkaido University, 3-1-1 Minato-cho, Hakodate 041-8611, Japan

ABSTRACT

The genus *Porphyra* includes several economically valuable species such as *P. yezoensis*, *P. tenera* and *P. pseudolinearis*. They are predominantly consumed and cultivated in East Asian countries such as Japan, Korea and China, and are regarded as a big commercial market today. In addition to the industrial importance, *P. yezoensis* is currently regarded as a feasible candidate for a model plant in marine sciences; therefore, there are a wide range of studies being undertaken: infrastructure arrangements, taxonomy, phylogeny, development, genetics, etc. Current situation and future perspectives in marine scientific studies using genus *Porphyra* will be summarized in this chapter.

INTRODUCTION

Marine algae are a diverse group of over eleven thousand species ranging in size from the microscopic unicellular plankton to macroscopic multicellular benthos. Marine macroalgae, collectively known as seaweed, are large enough to be seen with the naked eye, and they consist of the three major phyla, i.e., Rhodophyta (red algae), Chlomophyta (brown algae) and Chlorophyta (green algae). The members of marine macroalgae except Chlorophyta (green algae) have never gotten to land; most of them, if not all, have adapted themselves to marine circumstances and accomplished their unique evolution. Since survival strategy of marine macroalgae, i.e., photosynthetic mechanism, cell component structure, life cycle pattern, halophilism, etc., differs from those of land plants, they might have various

* Correspondence to: Naotsune Saga, Faculty of Fisheries Sciences, Hokkaido University, 3-1-1 Minato-cho, Hakodate 041-8611, Japan, Tel/Fax: +81-138-40-5533, E-mail: nsaga@fish.hokudai.ac.jp.

characteristics and biological functions peculiar to marine circumstances. It is desirable to establish some proper model organisms for efficient promotion of advanced studies of their own characters and special functions. In a recent review (Waaland et al., 2004), various kinds of taxa were proposed as candidate macroalgal models for genomics, e.g., the genus *Porphyla* as a Rhodophytic candidate, the genus *Ectocarpus* as a Chlomophytic candidate and the genus *Ulva* as a Chlorophytic candidate. Post-genome analysis is now main stream in modern biology using eukaryotic organisms including yeast, nematode, fly, mouse, human, etc. In plants, genome analyses had been done for several land plants, such as *Arabidopsis, Oryza* or *Physcomitrella* and unicellular microalgae, such as *Chlamydomonas*, *Cyanidioschyzon* or *Thalassiosira*. In addition, large scale-genome analysis of multicellular macroalgae such as *Ectocarpus* (Cock et al., 2010) and *Porphyra* is recently promoted in a world-wide level. It has already been organized some genome consortium for *Ectocarpus* in EC countries and *Porphyra* in Asian countries to which high amounts of national and/or international budget are invested.

The present status of infrastructure arrangements for advanced studies on *Porphyra* as model organisms in marine sciences will be introduced in this chapter, featuring the recent activities especially for *P. yezoensis*.

Porphyra as Model Organisms

The members of *Porphyra*, a genus of marine macroalgae, are distributed all over the world, and there are 133 species recognized in the world and 29 species recognized in Japan (Yoshida et al., 1995). *Porphyra* includes several economically valuable species, *P. yezoensis*, *P. tenera*, etc., which are the most valuable marine crops in the world. Their recent annual yields are about ten billion sheets, and recent annual sales of their products are estimated at multibillion U.S. dollars.

Porphyra has recently received great interest as model plants for fundamental and applied studies in marine sciences. *Porphyra* has a unique dimorphic life cycle, which consists of a leafy gametophytic generation and a filamentous sporophytic phase (Figure 1). Some species of *Porphyra* have an asexual cycle consisting of repetition of leafy gametophytes through monospores. It completes its life cycle in laboratory culture within a few months and has a small number of chromosomes (two to seven) in the genus *Porphyra* ($2.6\text{-}5.3 \times 10^8$bp) (Matsuyama-Serisawa et al., 2007) are in the same order of magnitude as those of *Arabidopsis thaliana*. Genetic analyses, including both classical and modern molecular studies, are in progress.

The members of our research group are performing the following subjects in order to make *Porphyra yezoensis* a sophisticated model organism: strain-preservation and cryo-preservation (Kuwano et al., 1994), establishment of mutant cell lines (Tomimatsu et al., 2005; Tomimatsu et al., 2006), transient gene expression systems (Fukuda et al., 2008; Mikami et al., 2009; Uji et al., 2010; Takahashi et al., 2010), genetic map (Park et al., 2007, 2008; Choi et al., 2008), EST (Nikaido et al., 2000; Asamizu et al., 2003), DNA macro/micro array (Kitade et al., 2008), etc.

The marine red alga *Porphyra yezoensis* has been proposed as a model plant for physiological and genetic studies in seaweed because of its biological and economical importance.

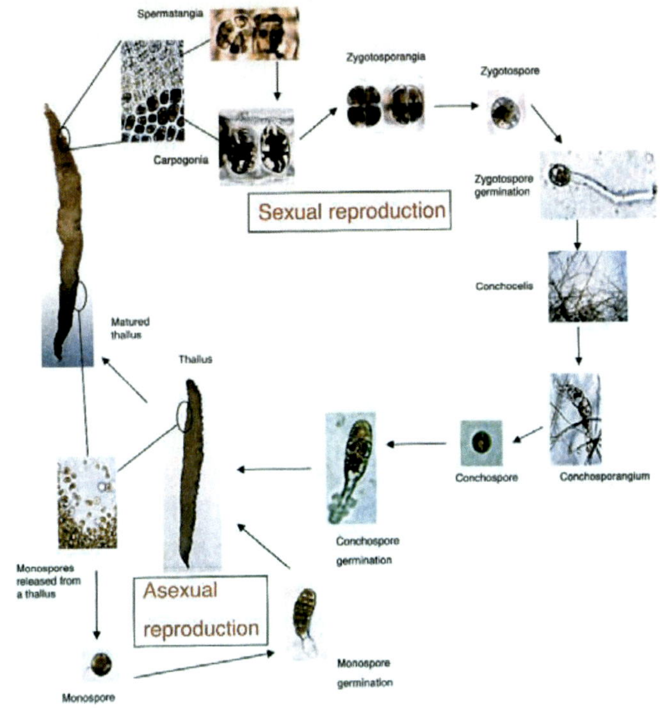

Figure 1. Life cycle of *Porphyra yezoensis*. It has a heteromorphic life cycle with an alternation between gametohytic phase and sporophytic phase. It propagates through both sexual and asexual modes of reproduction. In the asexual life cycle, monospores are released from leafy gametophytes and grow into the genetically identical offsprings.

However, the progress of molecular biological studies using gene transfection and genetic transformation systems has been hindered by difficulties in the expression of foreign genes in *P. yezoensis* cells. To overcome this situation, we developed transient gene expression systems to monitor gene expression in *P. yezoensis* cells using a particle bombardment. Since genes of *P. yezoensis* have GC-rich protein cording region, we first developed codon-optimized β-gluclonidase (GUS) reporter named PyGUS, which achieved efficient expression in *P. yezoensis* cells under the control of endogenous strong promoters such as *PyGAPDH* and *PyAct1* promoters (Fukuda et al., 2008; Takahashi et al., 2010). These studies indicated the importance of codon usage of foreign genes and use of endogenous promoters for efficient expression in *P. yezoensis* cells (Fukuda et al., 2008; Takahashi et al., 2010). Then these findings were applied for fluorescent reporters such as humanized AmCFP and plant-adapted sGFP, which enabled the visualization of subcellular localization of several proteins in *P. yezoensis* (Mikami et al., 2009; Uji et al., 2010). Establishment of the transient gene expression systems in *P. yezoensis* could facilitate biotechnological developments in this organism.

For generalizing *P. yezoensis* as a more sophisticated model, we are also trying to develop the genetic linkage map for this alga. As an initial step for this purpose, we established several pure lines for genetic analysis and examined genetic distances among those strains. These efforts resulted in identification of two appropriate strains (named TU-2 and KGJ strains), which can be readily used in molecular genetic studies. DNA-based genetic markers are efficient and robust tools for providing the high-resolution diagnostic traits required for detailed linkage analysis (Park et al., 2007). In general, co-dominant molecular markers can allow more precise analysis and provide a good wealth of information. Therefore, using cleaved amplified polymorphic sequence (CAPS) analysis, we developed several kinds of nuclear gene markers.

Taxonomy and Phylogeny

Bangiophyceae (Rhodophyta) is comprised of five orders including Bangiales (van den Hoek et al., 1995), and Bangiales contains only one family (Bangiaceae) with two genera, *Bangia* and *Porphyra* (Chihara, 1997). Because of the economic importance of *Porphyra*, extensive taxonomic studies have been performed in this genus (Kurogi, 1972; Conway et al., 1975; Lindstrom and Cole, 1992a, 1992b), and over 130 species have been described worldwide (Yoshida et al., 1997). The taxonomy of *Bangia* is complex, and many species are unified, but at least several species have remained (Cole and Sheath, 1990). Although some species of *Bangia* occur in fresh water, most species of *Bangia* and *Porphyra* grow in intertidal rocky substrata in seawater from subarctic to subtropical regions (Sheath and Cole, 1984; Cole and Sheath, 1990). Generic concepts divided into *Bangia* and *Porphyra* are based on the morphological characters of their macroscopic thalli; the thallus of *Bangia* represents a cylindrical form, while that of *Porphyra* is a leafy form. Kurogi (1972) classified *Porphyra* into three subgenera, *Diplastidia*, *Diploderma* and *Porphyra*, based on the number of cell layers (monostroma or distroma) and the plastid number per cell (monoplastid or diplastid). In general, species of *Porphyra* have been delineated by morphological and cytological features of the gametophytes (Chang and Zheng, 1960; Miura 1961, 1968; Kurogi, 1972; Oliveira and Cole, 1975; Abbott and Hollenberg, 1976; Cole and Sheath, 1990; Lindstrom and Cole, 1992a; Lindstrom and Cole, 1993; Notoya et al., 1993; Matsuo et al., 1994; Miyata and Kikuchi, 1997; Nelson et al., 1998; Yoshida, 1998).

Although *Bangia* had been considered to be a sister group to *Porphyra* (Garbary et al., 1980), recent molecular phylogenetic trees have contradicted the traditional placement of *Bangia* (Stiller and Waaland, 1993; Oliveira et al., 1995; Brodie et al., 1997; Woolcott and King, 1998). Based on sequences of nuclear small-subunit (SSU) rRNA genes, Oliveira et al. (1995) showed that the distromatic species, *Porphyra amplissima*, is more closely related to the monostromatic *P. leucosticta* than to the distromatic *P. miniata*, despite the fact that these species were previously placed in the subgenus *Diploderma*. Stiller and Waaland (1993) also suggested that the distromatic species *P. cuneiformis* appeared more closely related to the monostromatic *P. nereocystis* and *P. thuretii* than to the distromatic *P. scizophylla* on the basis of restriction fragment-length polymorphism analysis. However, no studies have been conducted to determine the molecular phylogeny of diplastidic species of *Porphyra*. Thus, molecular information on Bangiales is insufficient to date.

The SSU rRNA gene sequences of 59 strains of Bangiales and two strains of Erythropeltidales (outgroup) were obtained from the GeneBank database was used for construction of the phylogenetic trees (Table 1).

Table 1. List of strains used for molecular phylogenetic analysis

Genus /subgenus	Species	Accession no.	Reference	
	Bangia atropurpurea (Roth) C. Agardh	L36066	Oliveira et al. (1995)	*
	Bangia sp.	AF043353	Müller et al. (1998)	*
		AF043354	Müller et al. (1998)	*
		AF043355	Müller et al. (1998)	*
		AF043356	Müller et al. (1998)	*
		AF043357	Müller et al. (1998)	*
		AF043358	Müller et al. (1998)	*
		AF043359	Müller et al. (1998)	*
		AF043360	Müller et al. (1998)	*
		AF043361	Müller et al. (1998)	*
		AF043362	Müller et al. (1998)	*
		AF043363	Müller et al. (1998)	*
		AF043364	Müller et al. (1998)	*
		AF043365	Müller et al. (1998)	*
Porphyra (subgenus)				
	Porphyra acanthophora Oliveira Filho et Cole	L26197	Oliveira et al. (1995)	*
	P. dentata Kjellman	AB013183	Kunimoto et al. (1999)	*
	P. katadae Miura	AB013184	Kunimoto et al. (1999)	*
	P. leucosticta Thuret	L26199	Oliveira et al. (1995)	*
	P. lilliputiana Nelson, Knight et Hawkes	AF136424	Broom et al. (1999)	*
	P. pseudolinearis Ueda	AB013185	Kunimoto et al. (1999)	
		AB015793	Matsuyama-Serisawa et al. (2004)	*
		AF116913	Hong (unpublished)	*
	P. purpurea (Roth) C. Agardh	L26201	Oliveira et al. (1995)	*
	P. spiralis Oliveira Filho et Coll	L26177	Oliveira et al. (1995)	*
	P. suborbiculata Kjellman	AB013180	Kunimoto et al. (1999)	
		AB015796	Matsuyama-Serisawa et al. (2004)	*
	P. tenera Kjellman	AB013175	Kunimoto et al. (1999)	*
		AB013176	Kunimoto et al. (1999)	
		AB029879	Kunimoto et al. (unpublished)	

Table 1. (Continued)

Genus /subgenus	Species	Accession no.	Reference	
		AB029880	Kunimoto et al. (unpublished)	
		AB029881	Kunimoto et al. (unpublished)	
		AB029882	Kunimoto et al. (unpublished)	
		AB029883	Kunimoto et al. (unpublished)	
		AB029884	Kunimoto et al. (unpublished)	
	P. tenuipedalis Miura	AB015797	Matsuyama-Serisawa et al. (2004)	*
	P. umbilicalis (L.) J. Agardh	AB013179	Kunimoto et al. (1999)	*
		L26202	Oliveira et al. (1995)	
		L36049	Oliveira et al. (1995)	
	P. yezoensis Ueda	AB013177	Kunimoto et al. (1999)	*
		AB013178	Kunimoto et al. (1999)	
Diploderma (subgenus)				
	P. amplissima (Kjellman) Setchell et Hus	AB015791	Matsuyama-Serisawa et al. (2004)	*
		L36048	Oliveira et al. (1995)	*
	P. miniata (C. Agardh) C. Agardh	L26200	Oliveira et al. (1995)	*
	P. variegata (Kjellman) Kjellman	AB015792	Matsuyama-Serisawa et al. (2004)	*
Diplastidia (subgenus)				
	P. haitanensis Chang et Zheng	AB013181	Kunimoto et al. (1999)	
		AB015795	Matsuyama-Serisawa et al. (2004)	*
	P. onoi Ueda	AB015794	Matsuyama-Serisawa et al. (2004)	*
	Porphyra sp.	AB013182	Kunimoto et al. (1999)	*
		AF117239	Hong (unpublished)	*
		AF117306	Hong (unpublished)	*
		AF136420	Broom et al. (1999)	*
		AF136421	Broom et al. (1999)	*
		AF136422	Broom et al. (1999)	*
		AF136423	Broom et al. (1999)	*
		AF136426	Broom et al. (1999)	*
		AF136427	Broom et al. (1999)	*
		AF136428	Broom et al. (1999)	*
	Erythrocladia sp.	L26188	Oliveira et al. (1995)	*
	Erythrotrichia carmea (Dillwyn) J. Agardh	L26189	Ragan et al. (unpublished)	*

*Strains used for tracing of systematic characters

Three phylogenetic trees, the NJ tree, the most parsimonious (MP) tree, and ML tree, were constructed based on the SSU rRNA sequences of the 59 strains. MP analysis produced 54 parsimonious trees with 1,969 steps (consistency index: 0.53, retention index: 0.73). Among the three trees, the highest bootstrap values were confirmed in the NJ tree (Figure 2). In all trees, *Porphyra* sp. (AF117239) branched first, then *Bangia* sp. (AF043364) branched, and other Bangiales were largely divided into three clades.

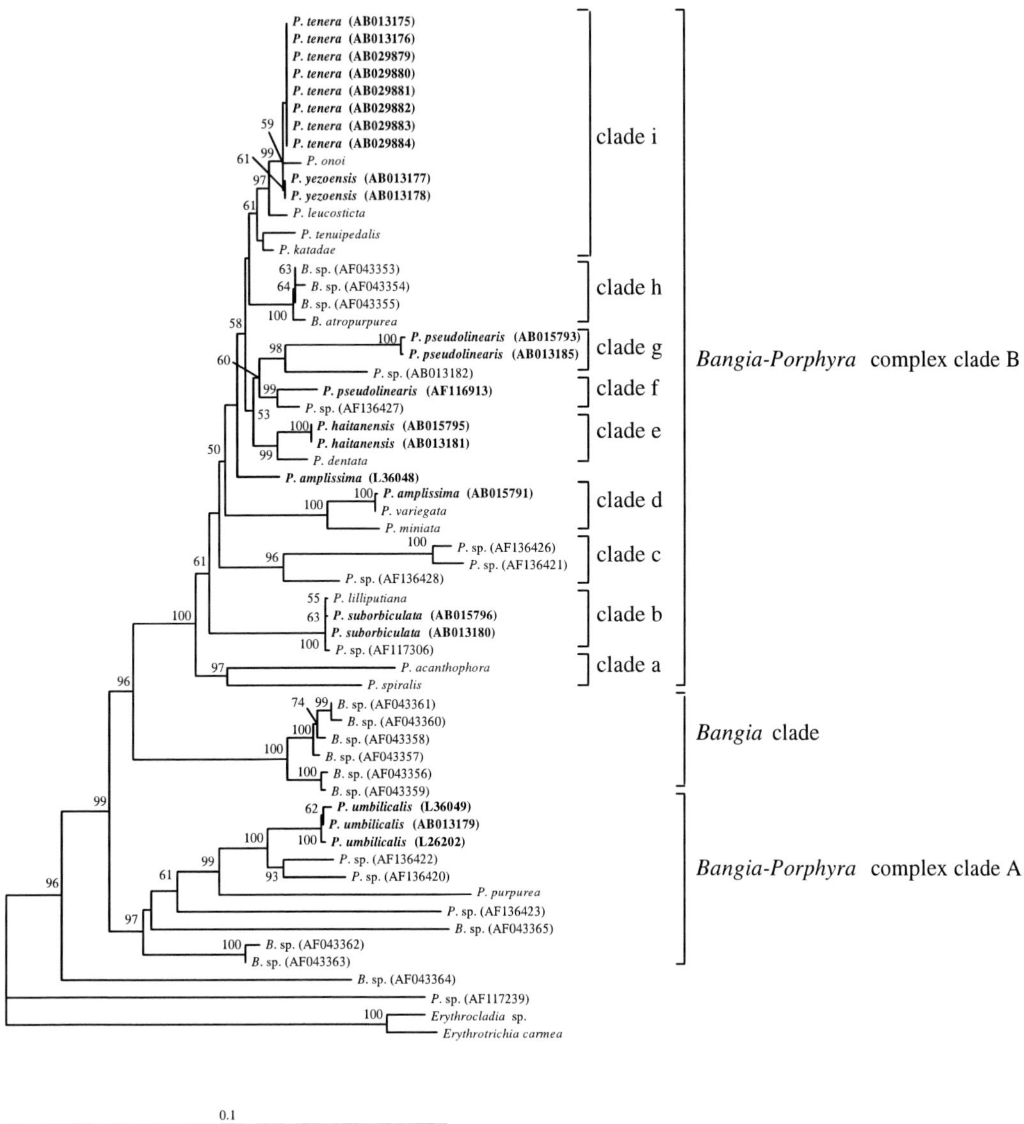

Figure 2. Phylogenetic tree of Bangiales inferred from sequences of nuclear SSU rRNA genes constructed using the neighbor-joining method. Scale bar indicates an evolutionary distance of 0.1 nucleotide substitution per position in the sequence. The numbers under the branches indicate bootstrap values greater than 50.

All strains of *Porphyra* except one (AF117239) were contained in the two clades with some strains of *Bangia*. Among the three clades, the *Bangia-Porphyra* complex clade A diverged first, then the *Bangia* clade diverged, and finally the *Bangia-Porphyra* complex clade B diverged. In the *Bangia-Porphyra* complex clade A, strains of *Bangia* branched first and then strains of *Porphyra* branched.

In the *Bangia* clade, only strains of *Bangia* branched. In the *Bangia-Porphyra* complex clade B, all distal branches (Figure 2, clades a–i) were divided similarly in all trees, but the ML tree formed a polytomy, and the order of branching in the MP tree was partially different from that in the NJ tree. The strains of *P. haitanensis*, *P. suborbiculata*, *P. tenera*, *P. umbilicalis*, and *P. yezoensis* were gathered within the same branches in each species, while strains of *P. amplissima* and *P. pseudolinearis* were divided into different branches.

In phylogenetic trees inferred from SSU rRNA sequences, it is clear that the 57 strains of Bangiales diverged largely into three clades, the *Bangia-Porphyra* complex clade A, *Bangia* clade, and *Bangia-Porphyra* complex clade B in all three trees (NJ, MP, and ML trees) (Figure 2). As strains of *Bangia* separated into the three clades, it is considered that the genus *Bangia* and *Porphyra* are not monophyletic and *Bangia* is not a sister group of *Porphyra*. The same context was also confirmed by Oliveira et al. (1995). It is interesting that strains of *P. amplissima* and *P. pseudolinearis* did not gather in one branch in each species. In those strains, further taxonomic investigations are thought to be necessary. Also, further studies will be required in the strains of unidentified species (*B.* sp. and *P.* sp.) of *Bangia* and *Porphyra* dealt with in the study by Matsuyama-Serisawa et al. (2004).

Tracing of Systematic Characters

A study on tracing of systematic characters based on molecular phylogeny in Bangiales is insufficient to date, but recently Matsuyama-Serisawa et al. (2004) reported the paper that performed the subject aforementioned.

Tracing of three systematic characters of a genus or subgenus were as follows. Character 1 (shape of gametophytes): Both cylindrical and leafy strains were included in *Bangia-Porphyra* complex clades A and B. In the *Bangia-Porphyra* complex clade B, cylindrical strains were gathered in clade c, and all other distal branches were composed of leafy strains. The *Bangia* clade was composed only of cylindrical strains. Thus, cylindrical strains separated the three clades. Character 2 (number of cell layers in gametophytes): Monostromatic strains appeared in *Bangia-Porphyra* complex clades A and B. Distromatic strains appeared only in the *Bangia-Porphyra* complex clade B and gathered in clade d. Character 3 (number of chloroplasts per cell in gametophytes): Most strains were judged to have one chloroplast per cell. Only two strains of diplastid appeared in clade e and clade i of the *Bangia-Porphyra* complex clade B.

Similarly, in each systematic character 4–15, which separate the species of *Bangia* and *Porphyra*, strains with the same character status appeared in various lineages as follows. Character 4: The thin form of thallus appeared in *Bangia-Porphyra* complex clades A and B. The thick form of thallus appeared in clades d–h and clade i of the *Bangia-Porphyra* complex clade B. Character 5: Most strains were judged to be pinkish-red to deep rose-pink. Nine strains with a reddish-brown to brown-red color appeared in seven positions of *Bangia-Porphyra* complex clades A and B. Character 6: *Porphyra* strains with marginal undulation appeared in clades a–e and clade i of the *Bangia-Porphyra* complex clade B (clade c is presumptive). Character 7: *Porphyra* strains with microscopic marginal dentation appeared in clade a, clade b, and clade e of the *Bangia-Porphyra* complex clade B. Character 8: Strains with three chromosomes appeared in clades a–d, clade h, and clade i of the *Bangia-Porphyra* complex clade B (clade c and clade h are presumptive). Strains with four chromosomes appeared in the *Bangia-Porphyra* complex clade A and in clade b, clades e–g, and clade i of the *Bangia-Porphyra* complex clade B. Strains with five chromosomes appeared in the *Bangia-Porphyra* complex clade A and clade e and clade i of the *Bangia-Porphyra* complex clade B. Character 9: All strains appearing in winter were contained in the *Bangia-Porphyra* complex clade B. Some strains appearing in summer were included in the *Bangia-Porphyra* complex clade A and clade d and clade i of the *Bangia-Porphyra* complex clade B. Character 10: Most strains were monoecious. Dioecious strains appeared in clades e–h of the *Bangia-Porphyra* complex clade B. Character 11: Strains with separated distribution of sexual cells by a vertical line were included in the *Bangia-Porphyra* complex clade A and clade d and clade i of the *Bangia-Porphyra* complex clade B. Strains with mixed distribution of sexual cells were contained in the *Bangia-Porphyra* complex clade B. Character 12: Strains with 128 of the Hus formula of spermatangia were included in the *Bangia-Porphyra* complex clade A, and strains with the three types of the Hus formula were mixed in the *Bangia-Porphyra* complex clade B. Character 13: Strains with 16 and 32 of the Hus formula of carposporangia were included in the *Bangia-Porphyra* complex clade A. Strains with three types of the Hus formula were mixed in the *Bangia-Porphyra* complex clade B. Character 14: Some strains in which asexual reproduction was confirmed appeared in the *Bangia-Porphyra* complex clade B. Character 15: Most strains formed conchospores. Strains forming unseparated conchospores or conchospores and monospores appeared in clade i or clade d, respectively.

Species of Bangiales have mainly been classified based on 15 systematic characters (Chang and Zheng, 1960; Miura, 1961, 1968; Kurogi, 1972; Oliveira and Cole, 1975; Abbott and Hollenberg, 1976; Cole and Sheath, 1990; Lindstrom and Cole, 1992a; Lindstrom and Cole, 1993; Notoya et al., 1993; Matsuo et al., 1994; Miyata and Kikuchi, 1997; Nelson et al., 1998; Yoshida, 1998), and the genus *Porphyra* was divided into three subgenera based on number of cell layers and chloroplasts per cell (Kurogi, 1972). Among the 15 systematic characters, the number of cell layers was reported not to reflect the molecular phylogeny of *Porphyra* (Stiller and Waaland, 1993; Oliveira et al., 1995). In the study by Matsuyama-Serisawa et al. (2004), from the cladistic tree of systematic character 1 (shape of gametophytes), which separates the genera of *Bangia* and *Porphyra*, it is considered that an evolution of gametophyte shape from cylindrical to leafy occurred at least twice and an evolution of the shape from leafy to cylindrical, once. From the cladistic tree of systematic character 2 (number of cell layers in gametophytes), which separates the subgenus *Diploderma* from the genus *Porphyra*, it is considered that an evolution of cell layers from monostromatic to distromatic occurred once, but monostromatic strains appeared in two lineages. The cladistic tree of systematic character 3 (number of chloroplasts per cell in gametophytes), which separates the subgenus *Diplastidia* from the genus *Porphyra*, suggests that an evolution of chloroplast number per cell from monoplastid to diplastid occurred twice in different positions. From the cladistic trees of systematic characters 4–15, which separate the species of *Bangia* and *Porphyra*, it seems that same status of the characters appeared in various lineages, although several assemblages were confirmed. Therefore, the 15 systematic characters used in the present study were concluded not to reflect the SSU rRNA gene phylogeny.

Life Cycle

We have been studying the development and regulation of life cycle in *Porphyra*. During these studies, we are particularly interested in molecular biology concerning the sexual cell differentiation and determination of generations. As mentioned above, there is the heteromorphic altermation of generations between a leafy gametophyte and a filamentous sporophyte in *Porphyra*. Using a scanning electron microscope (SEM), a transmission electron microscope (TEM) and a biochemical approach, a cytological survey of the vegetative cells of the gametophytic and sporophytic phases in *Porphyra* showed that two stages differ in a number of structural characteristics: form and growth (diffuse vs. apical), chloroplast number (one vs. many), pit plug (absent vs. present), vacuole (peripheral vs. central) and cell wall structure (xylan vs. cellulose). The morphogenetic modes of them are different, and this may be explained by different gene expression between two generations. Our interest is focused on the developmental stage- or generation phase-specific gene expression in *Porphyra*.

Genetics

Development of Mutant Strains

In *P. yezoensis*, pigment mutations have been generally used as genetic markers. Yan and Aruga (1997) produced several pigmentation mutants by treatment with N-methyl-N'-nitro-N-nitrosoguanidine (MNNG). Tomimatsu et al. (2006) developed a revised method of Yan and Aruga (1997) and succeeded in inducing a different kind of greenish mutant strain named MBG from TU-1 strain. Some of the mutants described above have been utilized in genetic analyses (Takahashi et al., 2008); however, detailed investigations at molecular level have not been reported to date. Therefore, disjunction between classical Mendelian genetics and molecular approach is a serious problem to overcome. In addition, in the case of *Porphyra* species, most of the mutants reported to date are the pigmentation mutants. Thus, development of other kinds of mutants with some physiological defects, such as auxotrophic or sexually sterile strains, would provide new insights in genetic analysis of red algae.

*Molecular Genetics*Cross-breeding of *P. yezoensis* has long been conducted for improvement of strains since the 1980s, and in the early days, pigmentation mutants were extensively used in crossing. Although pigment mutation provides an unequivocal character, which enables easy confirmation of heterozygosity by color, it requires a long waiting time because visible chimeric thalli usually appear several months after fertilization. To overcome this problem, our research group has developed a new rapid method for identification of cross fertilization by two CAPS markers for nuclear genes (Choi et al., 2008). Using these markers, we could examine heterozygosity of the conchocelis easily, and we can avoid the time-consuming process. Since *P. yezoensis* is a monoecious alga in which the female and male gametes are formed in one blade, isolation of fertilized zygotes requires the highly skilled technique. Therefore, contamination of self-fertilized conchocelis sometimes occurs during cross experiment, and the nuclear markers cannot distinguish them. This may lead to miscounting and fatal errors in statistical analysis in genetic studies. To solve this problem, we recently developed two CAPS markers for chloroplast and mitochondrial genomes. Using these markers, we can thoroughly exclude the self-fertilized conchocelis. In summary, molecular genetic analysis in *P. yezoensis* is still in the dawning; however, we have prepared the two absolutely necessary tools, strains and markers.

CONCLUSION

The arrangement for an infrastructure of advenced studies of *P. yezoensis* is a key to opening the door towards a new marine botanical world, which leads us stepping into the next phase of innovative studies of marine macro-algae especially the genus *Porphyra.*

ACKNOWLEDGMENT

I would like to thank Dr. Koji Mikami (Hokkaido University) and Mr. Ryo Hirata (Hokkaido University) for their valuable comments.

REFERENCES

Abbott, I. A. and Hollenberg, G. J. (1976). *Marine Algae of California*. Palo Alto: Stanford University Press.

Asamizu, E., Nakajima, M., Kitade, Y., Saga, N., Nakamura, Y. and Tabata, S. (2003). Comparison of RNA expression profiles between the two generations of *Porphyra yezoensis* (Rhodophyta), based on expressed sequence tag frequency analysis. *Journal of Phycology*, *39*, 923-930.

Brodie, J., Hayes, P. K., G. L. Barker, G. L. and Irvine, L. M. (1997). Speciation within the genus *Porphyra* (Bangiophycidae, Rhodophyta): evidence for a Pacific element within the Atlantic flora. *Phycologia*, *36* (*supplement*), 11.

Broom, J. E., Jones, W. A., Hill, D. F., Knight, G. A. and Nelson, W. A. (1999). Species recognition in New Zealand *Porphyra* using 18S rDNA sequencing. *Journal of Applied Phycology*, *11*, 421-428.

Chang, T. J. and Zheng, B. F. (1960). *Porphyra haitanensis*, new species of *Porphyra* from Fukien. *Acta Botanica Sinica*, *9*, 32-36 (in Chinese with English abstract).

Chihara, M. (1997). Rhodophyta. In M. Chihara (ed.), *Biology of Algal Diversity* (pp. 49-71). Tokyo: Uchida Rokakuho. (in Japanese).

Choi, S-J., Park, E-J., Endo, H., Kitade, Y. and Saga, N. (2008). Inheritance pattern of chloroplast and mitochondrial genomes in artificial hybrids of *Porphyra yezoensis* (Rhodophyta). *Fisheries Science, 74*, 822-829.

Cock, J.M., Sterck, L., Rouze, P., Scornet, D., Allen, A.E., Amoutzias, G., Anthouard, V., Artiguenave, F., Aury, J.M., Badger, J.H., Beszteri, B., Billiau, K., Bonnet, E., Bothwell, J.H., Bowler, C., Boyen, C., Brownlee, C., Carrano, C.J., Charrier, B., Cho, G.Y., Coelho, S.M., Collen, J., Corre, E., Da Silva, C., Delage, L., Delaroque, N., Dittami,

S.M., Doulbeau, S., Elias, M., Farnham, G., Gachon, C.M.M., Gschloessl, B., Heesch, S., Jabbari, K., Jubin, C., Kawai, H., Kimura, K., Kloareg, B., Kupper, F.C., Lang, D., Le Bail, A., Leblanc, C., Lerouge, P., Lohr, M., Lopez, P.J., Martens, C., Maumus, F., Michel, G., Miranda-Saavedra, D., Morales, J., Moreau, H., Motomura, T., Nagasato, C., Napoli, C.A., Nelson, D.R., Nyvall-Collen, P., Peters, A.F., Pommier, C., Potin, P., Poulain, J., Quesneville, H., Read, B., Rensing, S.A., Ritter, A., Rousvoal, S., Samanta, M., Samson, G., Schroeder, D.C., Segurens, B., Strittmatter, M., Tonon, T., Tregear, J. W., Valentin, K., von Dassow, P., Yamagishi, T., Van de Peer, Y. and Wincker, P. (2010) The *Ectocarpus* genome and the independent evolution of multicellularity in brown algae. *Nature*, *465*, 617-621

Cole, K. M. and Sheath, R. G. (1990). *Biology of the Red Algae*. Cambridge: Cambridge University Press.

Conway, E., Mumford, Jr., T. F. and Scagel, R. F. (1975). The genus *Porphyra* in British Columbia and Washington. *Syesis*, *8*, 185-244.

Fukuda, S., Mikami, K., Uji, T., Park, E. J., Ohba, T., Asada, K., Kitade, Y., Endo, H., Kato, I. and Saga, N. (2008). Factors influencing efficiency of transient gene expression in the red macrophyte *Porphyra yezoensis*. *Plant Science*, *174*, 329-339.

Garbary, D. J., Hansen, G. I. and Scagel, R. F. (1980). A revised classification of the Bangiophyceae (Rhodophyta). *Nova Hedwigia*, *33*, 145-166.

Kitade, Y., Asamizu, E., Fukuda, S., Nakajima, M., Ootsuka, S., Endo, E., Tabata, S. and Saga, N. (2008). Identification of genes preferentially expressed during asexual sporulation in *Porphyra yezoensis* gametophytes (Bangiales, Rhodophyta). *Journal of Phycology*, *44*, 113-123.

Kunimoto, M., Kito, H., Yamamoto, Y., Cheney, D. P., Kaminishi, Y. and Mizukami, Y. (1999). Discrimination of *Porphyra* species based on small subunit ribosomal RNA gene sequence. *Journal of Applied Phycology*, *11*, 203-209.

Kurogi, M. (1972). Systematics of *Porphyra* in Japan. In I. A. Abbott, and M. Kurogi (eds.), *Contributions to the Systematics of Benthic Marine Algae of the North Pacific* (pp. 162-192). Kobe: Japan Society of Phycology.

Kuwano, K., Aruga, Y. and Saga, N. (1994). Cryopreservation of the conchocelis of *Porphyra* (Rhodophyta) by applying simple pre-freezing system. *Journal of Phycology*, *30*, 566-570.

Lindstrom, S. C. and Cole, K. M. (1992a). Relationships between some North Atlantic and North Pacific species of *Porphyra* (Bangiales, Rhodophyta): evidence from isozymes, morphology, and chromosomes. *Canadian Journal of Botany*, *70*, 1355-1363.

Lindstrom, S. C. and Cole, K. M. (1992b). A revision of the species of *Porphyra* (Rhodophyta: Bangiales) occurring in British Columbia and adjacent waters. *Canadian Journal of Botany*, *70*, 2066-2073.

Lindstrom, S. C. and Cole, K. M. (1993). The systematics of *Porphyra*: character evolution in closely related species. *Hydrobiologia*, *260/261*, 151-157.

Matsuo, M, Notoya, M. and Aruga, Y. (1994). Life history of *Porphyra suborbiculata* Kjellman (Bangiales, Rhodophyta) in culture. *La Mer*, *32*, 57-63 (in Japanese with English summary).

Matsuyama-Serisawa, K., Yamamoto, M., Fujishita, M., Endo, H., Serisawa, Y., Tabata, S., Kawano, S. and Saga, N. (2007) DNA content of the cell nucleus in the macroalga *Porphyra yezoensis* (Rhodophyta). *Fisheries Science*, *73*, 738-740.

Matsuyama-Serisawa, K., Yamazaki, S., Kitade, Y., Serisawa, Y., Kuwano, K. and Saga, N. (2004). Tracing of systematic characters of Bangiales (Rhodophyta) based on molecular phylogeny inferred from sequences of nuclear small subunit rRNA genes. *Aquaculture Science, 52*, 185-198.

Mikami, K., Uji, T., Takahashi, M., Yasui, H. and Saga, N. (2009) Visualization of phosphoinositides via the development of the transient expression system of a cyan fluorescent protein in the red alga *Porphyra yezoensis*. *Marine Biotechnology*, *11*, 563-569.

Miura, A. (1961). A new species of *Porphyra* and conchocelis-phase in nature. *Journal of the Tokyo University of Fisheries*, *47*, 305-311.

Miura, A. (1968). *Porphyra katadae*, a new species from Japanese coast. *Journal of the Tokyo University of Fisheries*, *54*, 55–59.

Miyata, M. and Kikuchi, N. (1997). Taxonomic study of *Bangia* and *Porphyra* (Bangiaceae, Rhodophyta) from Boso Peninsula, Japan. *Natural History Research*, *3* (*special issue*), 19-46.

Müller, K. M., Sheath, R. G., Vis, M. L., Crease, T. J. and Cole, K. M. (1998). Biogeography and systematics of *Bangia* (Bangiales, Rhodophyta) based on the Rubisco spacer, *rbc*L gene and 18S rRNA gene sequences and morphometric analyses. 1. North America. *Phycologia*, *37*, 195-207.

Nelson W. A., Knight, G. A. and Hawkes, M. W. (1998). *Porphyra lilliputiana* sp. nov. (Bangiales, Rhodophyta): a diminutive New Zealand endemic with novel reproductive biology. *Phycological Research*, *46*, 57-61.

Nikaido, I., Asamizu, E., Nakajima, M., Nakamura, Y., Saga N. and Tabata, S. (2000). Generation of 10, 154 expressed sequence tags from a leafy gametophyte of a marine red alga, *Porphyra yezoensis*. *DNA Research, 7*, 223-227.

Notoya, M., Kikuchi, N., Aruga, Y. and Miura, A. (1993). Culture studies of four species of *Porphyra* (Rhodophyta) from Japan. *Fisheries Science*, *59*, 431-436.

Oliveira, F. E. C. and Cole, J. (1975). The genus *Porphyra* C. Ag. (Rhodophyta, Bangiales) in the American south Atlantic. I. Brazilian species. *Botanica Marina*, *18*, 191–197.

Oliveira M. C., Kuniawan, J., Bird, C. J., Rice, E. L., Murphy, C. A., Singh, R. K., Gutell, R. R. and Ragan, M. A. (1995). A preliminary investigation of the order Bangiales (Bangiophycidae, Rhodophyta) based on sequences of nuclear small-subunit rRNA genes. *Phycological Research, 43*, 71-79.

Park, E-J., Fukuda, S., Endo, H., Kitade, Y. and Saga, N. (2007). Genetic polymorphism within *Porphyra yezoensis* (Bangiales, Rhodophyta) and related species from Japan and Korea detected by cleaved amplified polymorphic sequence analysis. *European Journal of Phycology, 42*, 29-40.

Park, E-J., Fukuda, S., Endo, H., Kitade, Y. and Saga, N. (2007). Identification of cross-fertilized conchocelis using cleaved amplified polymorphic sequence markers in cross-experiments of *Porphyra yezoensis* (Bangiales, Rhodophyta). *Journal of Phycology*, *44*, 331-334.

Peters, A. F., Marie, D., Scornet, D., Kloareg, B. and Cock, J. M. (2004). Proposal of *Ectocarpus siliculosus* (Ectocarpales, Phaeophyceae) as a model organism for brown algal genetics and genomics. *Journal of Phycology, 40*, 1079-1088.

Saga, N. and Kitade, Y. (2002). *Porphyra*: A model plant in marine sciences. *Fisheries Science, 68* (supplement), 1075-1078.

Sheath, R. G. and Cole, K. M. (1984). Systematics of *Bangia* (Rhodophyta) in North America. I. Biogeographic trends in morphology. *Phycologia*, *23*, 383-396.

Stiller, J. W. and Waaland, J. R. (1993). Molecular analysis reveals cryptic diversity in *Porphyra* (Rhodophyta). *Journal of Phycology*, *29*, 506-517.

Takahashi, M., Tomimatsu, R., Endo, H., Kitade, Y., Mizuta, H. and Saga, N. (2008). Phenotypic and genetic characterization of two pigmentation mutant strains of a marine red alga *Porphyra yezoensis* (Bangiales, Rhodophyta). *Fish Genetics and Breeding Science, 37*, 19-27 (in Japanese with English abstract).

Takahashi, M., Uji, T., Saga, N. and Mikami, K. (2010). Isolation and regeneration of transiently transformed protoplasts from gametophytic blades of the marine red alga *Porphyra yezoensis*. *Electronic Journal of Biotechnology* [online], vol. 13, no. 2, ISSN 0717-3458.

Tomimatsu, R., Takahashi, M., Endo, H., Kitade, Y., Mizuta, H. and Saga, N. (2005). A simple method for inducing mutants in *Porphyra yezoensis* (Bangiales, Rhodophyta). *Fish Genetics and Breeding Science, 34*, 123-128 (in Japanese with English abstract).

Tomimatsu, R., Takahashi, M., Endo, H., Kitade, Y., Yasui, H. and Saga, N. (2006). Induction and characterization of a brilliant green mutant in a marine red alga *Porphyra yezoensis* (Bangiales, Rhodophyta). *Fish Genetics and Breeding Science, 36*, 43-47.

Uji, T., Takahashi, M., Saga, N. and Mikami, K. (2010) Visualization of nuclear localization of transcription factors with cyan and green fluorescent proteins in the red alga *Porphyra yezoensis*. *Marine Biotechnology*, *12*, 150-159.

van den Hoek, C., Mann, D. G. and Jahns, H. M. (1995). Rhodophyta (red algae). In C. van den Hoek, D. G. Mann, and H. M. Jahns (eds.), *Algae: An Introduction to Phycology* (pp. 48–101). Cambridge: Cambridge University Press.

Waaland, J. R., Stiller, J. W. and Cheney, D. P. (2004). Macroalgal candidates for genomics. *Journal of Phycology*, *40*, 26-33.

Woolcott, G. W. and King, R. J. (1998). *Porphyra* and *Bangia* (Bangiaceae, Rhodophyta) in warm temperate waters of eastern Australia: Morphological and molecular analyses. *Phycological Research*, *46*, 111–123.

Yamazaki, A., Nakanishi, K. and Saga, N. (1998). Axenic tissue culture and morphogenesis in *Porphyra yezoensis* (Bangiales, Rhodophyta). *Journal of Phycology, 34*, 1082-1087.

Yan, X. H. and Aruga, Y. (1997). Induction of pigment mutants by treatment monospore germlings with NNG in *Porphyra yezoensis* Ueda (Bangiales, Rhodophyta). *Algae, 12*, 39-52.

Yoshida, T. (1998). *Marine Algae of Japan*. Tokyo: Uchida Rokakuho (in Japanese).

Yoshida, T., Notoya, N., Kikuchi, N. and Miyata, M. (1997). Catalogue of species of *Porphyra* in the world, with special reference to the type locality and bibliography. *Natural History Research, 3* (*special issue*), 5-18.

Yoshida, T., Yoshinaga, K. and Nakajima, Y. (1995). Check list of marine algae of Japan (revised in 1995). *Japanese Journal of Phycology, 43*, 115-171 (in Japanese with English abstract).

In: Porphyra yezoensis
Editor: Koji Mikami

ISBN 978-1-61122-304-0

Chapter 2

MONOSPORE-BASED ASEXUAL LIFE CYCLE IN *PORPHYRA YEZOENSIS*

Koji Mikami[1*]***, Lin Li*** [2 †] ***and Megumu Takahashi*** [2]

[1] Faculty of Fisheries Sciences, Hokkaido University, 3-1-1 Minato-cho, Hakodate 041-8611, Japan

[2] Graduate School of Fisheries Sciences, Hokkaido University, 3-1-1 Minato-cho, Hakodate 041-8611, Japan

ABSTRACT

Porphyra yezoensis is a marine red macroalga included in Bangiophycideae. It exhibits a complex life cycle that consists of sexual and asexual propagation. The asexual life cycle is based on production and growth of a haploid spore, a monospore, produced in haploid gametophytes. Knowledge about the process of the asexual life cycle in *P. yezoensis* has recently been accumulating, indicating that environmental factors such as light and extracellular Ca^{2+} content are critical for activation of calcium-dependent and phosphoinositide-mediated signal transduction pathways during the production and development of asexual monospores. Although the mechanisms regulating the asexual life cycle are largely unknown, discussions about its advantages as a survival strategy in Bangiophycean red algae are possible. Because of the aflagellate nature of red algal cells, syngamy (fertilization) of male and female gametes is infrequent compared with green and brown algae whose male spores have flagella that allow them to swim toward female gametes. Thus, we propose that asexual monospores are evolutionally employed for the production of a large number of male and female gametes, thus increasing the physical contact required for frequent syngamy in Bangiophycean red algae. This system is different from the spore-based asexual life cycles of filamentous fungi, in which asymmetric division of stem cells occurs and also from the spore-based sexual triphasic life cycle of Florideophycean red algae, in which an additional diploid phase results in the production of multiple diploid spores that develop into a number of sporophytes. That

---

* Correspondence to: Koji Mikami, Faculty of Fisheries Sciences, Hokkaido University, 3-1-1 Minato-cho, Hakodate 041-8611, Japan Tel/Fax: +81-138-40-8899, E-mail: komikami@fish.

† Present address: College of Fisheries and Life Science, Shanghai Ocean University, 999 Huchenghuan Road, Lingang New City,Shanghai 201306, People's Republic of China.

is, the monospore-based asexual life cycle found in Bangiophycean red algae is a unique strategy in terms of the increased success rate of syngamy.

INTRODUCTION

Establishment of a particular style of life cycle is a basic survival strategy, and thus, in multicellular eukaryotes, a wide variety of different life cycles are observed. Land and marine plants generally exhibit a biphasic sexual life cycle with morphological differences between gametophytes and sporophytes (Mable and Otto, 1998; Coelho et al., 2007). At present, sexual life cycles are classified into three types, according to the relative independence of the sporophyte and gametophyte phases: in ferns and flowering plants, a sporophyte-dominant (diplontic) cycle with a highly reduced gametophyte phase; in bryophytes, a gametophyte-dominant (haplontic) cycle with a parasitic sporophyte phase; and in marine macroalgae, a haploid-diploid (haplo-diplontic) cycle with reciprocal appearance of separate free-living sporophytes and gametophytes (Mable and Otto, 1998; Coelho et al., 2007). In these life cycles, meiosis in the diploid sporophyte produces a haploid gametophyte, while syngamy (fertilization) of male and female gametes restores the diploid sporophytic genome.

In addition to the sexual life cycles mentioned above, asexual life cycles also exist. For example, in marine macroalgae, free-living haploid gametophytes often produce asexual spores that develop into haploid gametophytic clones by mitotic cell division without ploidy change (Magne, 1991). Although it is well known that algae and fungi have an asexual life cycle in which asexual spores are produced and develop to gametophytes, asexual propagation in multicellular eukaryotes has so far only been studied extensively in filamentous fungi such as *Aspergilius nidulans* and *Neurospora crassa* (Springer, 1993; Adams et al., 1998; Osherov and May, 2001). Thus, the mechanisms regulating the asexual life cycle in algae are still largely unknown, mainly due to the lack of a model for investigation.

Porphyra yezoensis, a red macroalga included in Bangiophycideae (Rodophyta), has recently received considerable attention as a promising model macroalga for physiological and molecular biological studies of marine red algae. The model largely depends on the establishment of a laboratory culture system through which the haploid-diploid life cycle can be completed within a short period (Saga and Kitade, 2002). As with many Bangiophycean red algae, asexual spores, so-called monospores, are produced in monosporangia typically occurring at the marginal region of the gametophyte (Miura, 1985). Thus, *P. yezoensis* undergoes a monospore-based asexual life cycle, although its biphasic sexual life cycle, which consists of morphologically distinct macroscopic gametophytic blades in winter and microscopic sporophytic filaments in summer, is predominant (Kurogi, 1961; Miura, 1985).

Despite an accumulation of morphological and cytological observations, the mechanisms regulating sexual and asexual life cycles in red macroalgae are still unknown. Nevertheless, information regarding the regulatory mechanisms of the asexual life cycle in *P. yezoensis* has been accumulating compared with knowledge of the sexual life cycle. This chapter summarizes recent studies on the monospore-based asexual cycle in *P. yezoensis* and discusses the physiological significance of asexual spore formation by comparison of this life cycle with that of filamentous fungi and asexual spore-lacking *Porphyra* species.

COMPLEX LIFE CYCLES IN *P. YEZOENSIS*

Sexual Life Cycle

Historically, the leafy gametophytic thallus and filamentous sporophyte of *Porphyra* were identified as different algal species, with the latter being given the name *Conchocelis rosa*, a shell-boring organism (Batters, 1892). However, Dr. Kathleen Drew discovered that the filamentous conchocelis is a sporophytic stage in the life history of *P. umbilicalis* (Drew, 1949); subsequently, an identical link between the concocelis and thallus was also established in *P. tenera* (Kurogi, 1953; Tseng and Chang, 1955). It was, therefore, concluded that the life cycle of *Porphyra* is heteromorphic, consisting of a leafy thallus (gametophyte) and a filamentous conchocelis (sporophyte).

P. yezoensis represents the same heteromorphic cycle, a typical haploid-diploid cycle, as shown in Figure 1. In spring, sexual organs, spermatangium and carpogonium, develop in a mosaic on haploid gametophytic blades to produce respective spermatium and carpogonium. Syngamy (fertilization) between the two then follows resulting in diploid carposporangium, in which carpospores are produced via mitosis. Carpospores are then released and adhere to the surface of rocks and shells before growing into conchocelis. Next, a swollen sporophyte, known as a conchosporangium, develops on mature conchocelis, producing conchospores that are released and grow into leafy gametophytes in winter.

Meiosis truly occurs between the formation and first cell division of conchospores. Nuclear morphological studies based on microscopic observations suggest that meiosis in *P. yezoensis* occurs at conchospore formation (Migita 1967; Kito, 1974), although karyotype analysis (Ma and Miura, 1984; Wang et al., 2006) and cytogenetic analysis with color mutants (Ohme et al., 1986; Ohme and Miura, 1988) suggest that meiosis is completed when conchospores are germinated after release from conchosporangium. Identification of regulator molecules of conchospore production and analysis of their activation modes during development are, therefore, necessary to determine when exactly meiosis occurs.

Asexual Life Cycle

In winter, free-living haploid gametophytes of marine macroalgae produce asexual spores that develop into haploid gametophytic clones by mitotic cell division without ploidy change (Magne, 1991). Among plants, such asexual propagation is a unique survival strategy found in marine multicellular algae, increasing the number of gametophytic clones via the growth of asexual haploid spores. As with most Bangiophycean red algae, *P. yezoensis* undergoes a monospore-based asexual life cycle (Figure 1), in which monospores are produced in and released from monosporangia (Figure 2A) before migrating while undergoing morphological change. They then adhere to the substratum and divide asymmetrically to produce two different vegetative and rhizoid cells (Figure 2B) (Guiry, 1990; Pickett-Heaps et al., 2001; Ackland et al., 2007; Li et al., 2008a,b; Mikami, 2010).

Below, we have summarized current knowledge on each of the above topics and discuss the biological significance of the asexual life cycle in *P. yezoensis*.

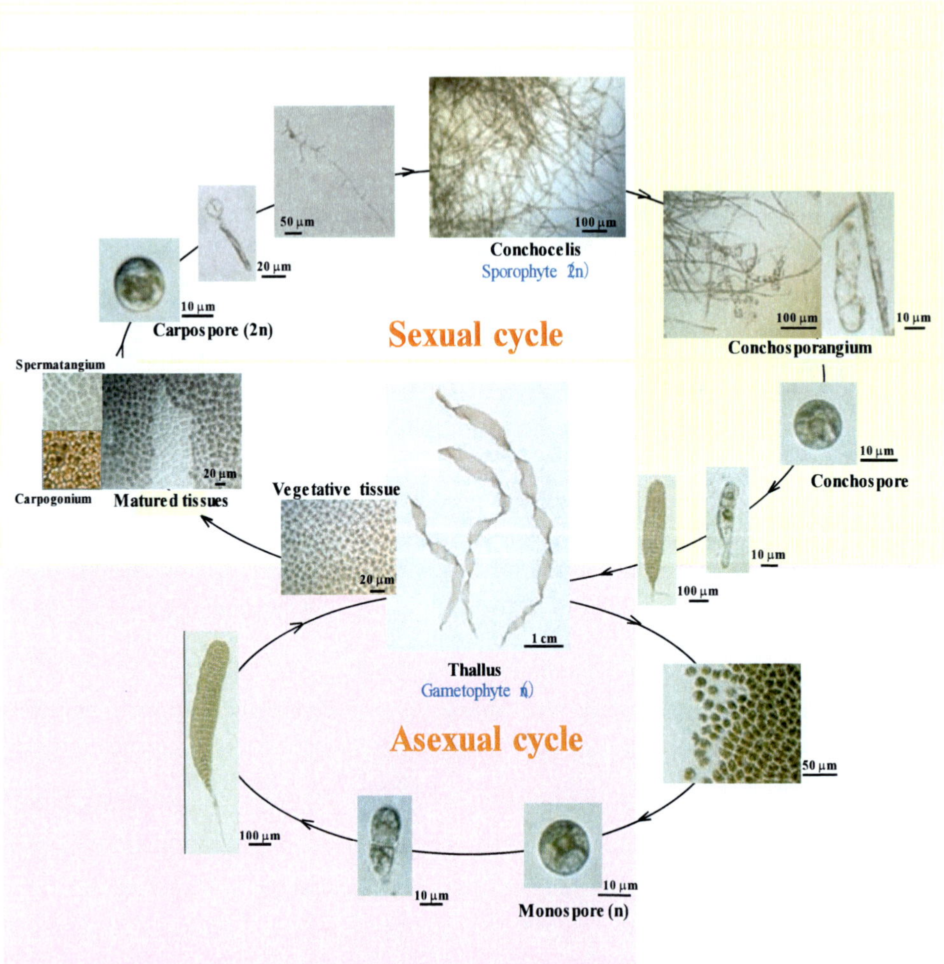

Figure 1. Life cycle of *P. yezoensis.* The sexual cycle belongs to a haploid-diploid type consisting of separate free-living leafy gametophytes and filamentous sporophytes. On mature gametophytes, haploid male and female gametes develop and are fertilized to produce diploid carporangia in which carpospores are produced mitotically. After discharge from the gametophyte, the carpospores are germinated into a filamentous conchocelis. The branch of the conchocelis sometimes develops into a swelling conchosporangium in which conchospores are produced. It is not clear whether the nuclear phase of conchospore development is haploid or diploid, since it is unknown when meiosis is completed during conchospore production and development. Released conchospores finally develop into a haploid gametophyte. In contrast, the asexual cycle depends on the production of haploid monospores from the gametophyte. Released monospores directly develop into a haploid gametophyte via mitosis, resulting in the production of genetic clones of the parent gametophyte.

Production and Discharge of Asexual Monospores

Asexual monospore production requires the formation of monosporangia, in which vegetative cells develop into small red-pigmented monospores (Figure 2Ab; Li et al., 2008a,b); these typically occur at the marginal region of the gametophyte in *P. yezoensis*

(Miura, 1985). Because the origin of asexual monospores can be traced back to 1,200 MYA based on fossil records of monospores from an ancient Bangiophycean red alga (Butterfield, 2000), research into the asexual life cycle of red algae could provide novel information regarding the origin and evolution of the mechanisms regulating the eukaryotic life cycle. However, to date, little is known about the mechanisms regulating the production and discharge of monospores from gametophytes in *P. yezoensis*, except for the effects of water temperature and light conditions on monospore release (Li, 1984; Kitade et al., 1998).

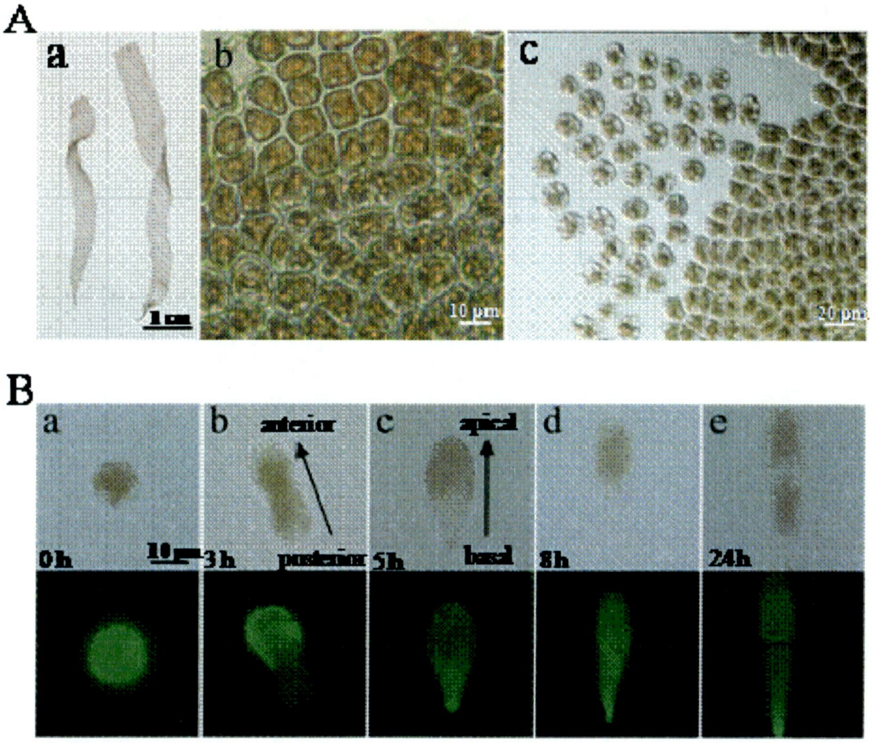

Figure 2. Production, discharge and early development of asexual monospores from *P. yezoensis.* (A) Production and discharge of monospores. (a) Haploid leafy gametophyte (thallus) of a laboratory-cultured *P. yezoensis* strain TU-1. (b) Magnified view of monosporangia and vegetative cells. Red colored and chloroplast-rich cells in monosporangia (left upper part), which are different from vegetative cells, are pre-monospores before discharge from gametophytes. (c) Discharge of unicellular monospores from a multicellular gametophyte. The developmental process of monospores has been described previously Li et al. (2008a) and Mikami (2010). (B) Morphological changes and F-actin distribution during early development of discharged monospores. F-actin was stained with Alex Flour 488 phalloidin. Upper and lower photographs show bright-field and F-actin fluorescent images. After discharge, monospores show a spherical form with F-actin bundles in the cell (a) before changing to a tapered form for migration with the establishment of the anterior-posterior axis and the asymmetrical distribution of F-actin at the leading edge (b). The end of migration triggers axis conversion for the establishment of an apical-basal axis in the upright form of monospores in which F-actin accumulates at the bottom and the chloroplast locates in the upper part (c). The upright monospore then elongates (d) and divides asymmetrically to produce a thallus and rhizoid cells by symmetrical cleaving of the chloroplast (e). After becoming upright, F-actin is continually localized at the bottom (c-e).

Extracellular Ca^{2+} Concentration Affects the Production and Discharge of Monospores

As shown in Figure 3A, growth inhibition was observed when the synthetic medium ASPMT1 (Takahashi et al., 2010), a modified version of ASP_{12} (Provasoli et al., 1957; Provasoli, 1963), was used for cultivation of gametophytes in comparison with growth in ESL (enriched SEALIFE) medium, which is routinely employed for laboratory culture of *P. yezoensis* (Kitade et al., 2002). In this case, vegetative cells of ASPMT1-cultured gametophytes assumed a round shape and deep red color (Figure 3A, center) and exhibited increased discharge of monospores relative to ESL-cultured gametophytes (Figure 3A, bottom). Importantly, these characteristics of vegetative cells resemble those of naturally occurring in monosporangia (Figure 2A), indicating that differences in the chemical composition of ASPMT1 and ESL are responsible for the enhanced production and discharge of monospores.

Measurement of the chemical composition of ESL using a polarized Zeeman atomic absorption spectrophotometer indicated that the concentration of Ca^{2+} in ESL is three times greater than that in ASPMT1, although the concentration of Mg^{2+} and K^{+} were equal between the two media (Takahashi et al., 2010). In addition, cultivation in two modified versions of ASPMT1 with differing $CaCl_2$ concentrations, two times greater (800 mg/l; ASPMT3) or half (200 mg/l; ASPMT4) of the original concentration of 400 mg/l, indicated normal growth and monospore discharge from gametophytes cultured in ASPMT3 and red coloration and extensive monospore discharge in ASPMT4-cultured gametophytes (Figure 3A; Takahashi et al., 2010). Therefore, the extracellular concentration of Ca^{2+} has an important influence on the asexual reproductive life cycle, controlling monospore production and discharge in *P. yezoensis*.

Critical role of photosynthesis-dependent Ca^{2+} influx in the production and discharge of monospores

It is well known that the discharge of monospores is light dependent in Bangiophycean red algae such as *P. yezoensis* and *Bangia atropurpurea* (Charnofsky et al., 1982; Li, 1984). Our previous experiments indicated that when gametophytes of *P. yezoensis* were treated for three days with 3-(3,4-dichlorophenyl)-1,1-dimethylurea (DCMU), an inhibitor of electron transport on the acceptor side of photosystem II, the number of discharged monospores decreased in a concentration-dependent manner (Takahashi et al., 2010). Thus, photosynthetic activity is an important factor regulating the asexual life cycle in *P. yezoensis*.

Determining the relationship between extracellular Ca^{2+} concentration and photosynthetic activity is, therefore, important. When extracellular Ca^{2+} was reduced by adding EGTA to ESL or by using ASPMT1 and ASPMT4 for culture (see Figure 3A, bottom), discharge of monospores was enhanced (Takahashi et al., 2010). Moreover, an artificial influx of Ca^{2+} with 1 μM calcium ionophore A23187 in the absence of illumination resulted in an enhancement of monospore discharge, although gametophytes cultured in the dark without A23187 did not produce monospores (Takahashi et al., 2010). Therefore, extracellular Ca^{2+} influx into vegetative cells is photosynthesis-dependent and a critical factor directing the asexual life cycle in *P. yezoensis*.

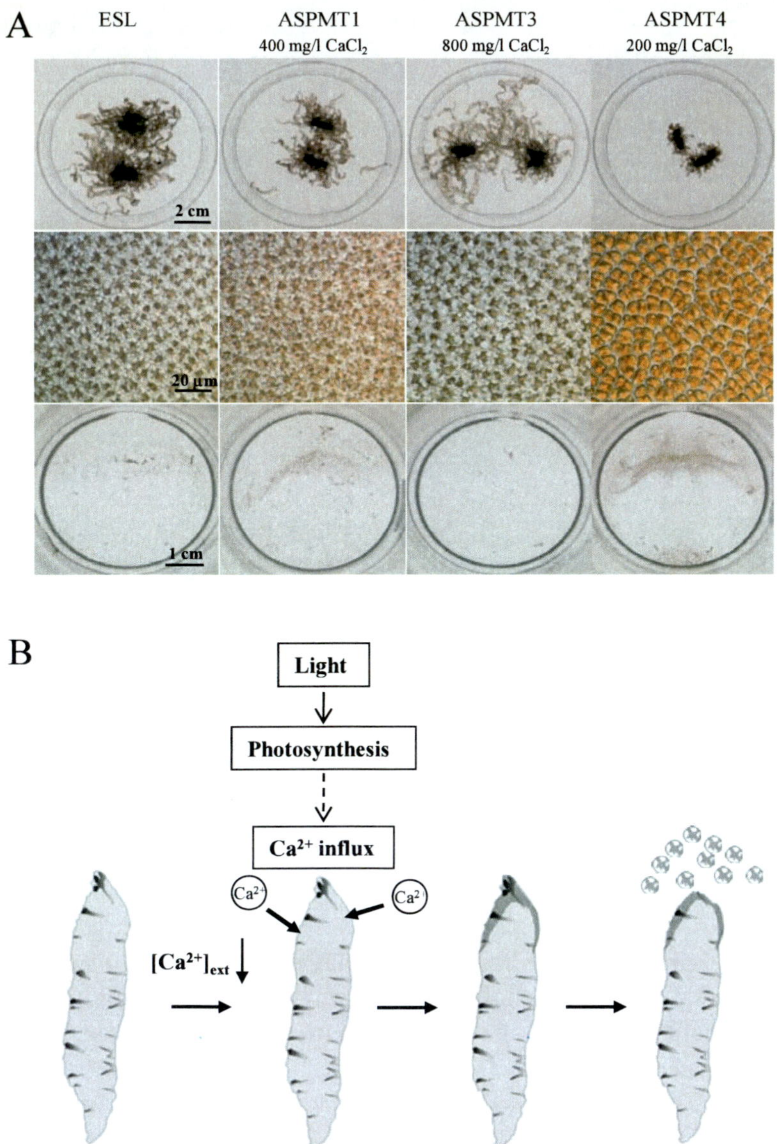

Figure 3. Effect of extracellular Ca^{2+} content on the production and discharge of monospores. (A) Comparison of growth, morphology of vegetative cells and discharge of monospores among gametophytes cultured in ESL, ASPMT1, ASPMT3 or ASPMT4. Gametophytes were cultured separately in each medium at 15°C under 60 μmol photons m^{-2} s^{-1} with a photocycle of 10 h light and 14 h dark for four weeks. Growth of gametophytes (upper panel), a microscopic view of vegetative cells (middle) and monospore accumulation in the bottom of the culture flasks (lower) are shown. Concentrations of $CaCl_2$ in ASPMT1, ASPMT3 and ASPMT4 are also indicated above each photo. (B) A model of the involvement of extracellular Ca^{2+} influx in the production and discharge of asexual monospores in *P. yezoensis* gametophytes. When the extracellular Ca^{2+} content decreases, the formation of monosporangia is stimulated at the tip region of gametophytes. This is followed by discharge of monospores, during which photosynthesis activated by illumination accelerates extracellular Ca^{2+} influx, promoting the formation of monosporangia. The area containing monosporangia is schematically indicated by the highlighted gray region.

Based on these findings, a mechanism for the formation and discharge of monospores was hypothesized, as shown in Figure 3B. Briefly, a reduction in extracellular Ca^{2+} concentration stimulates the formation of monosporangia at the edge of gametophytes in which photosynthesis-dependent extracellular Ca^{2+} influx is indispensable. Monosporangia then release monospores autonomously or in a Ca^{2+} influx-dependent manner. To clarify this possibility, identification of the nature of pumps or transporters of Ca^{2+}, whose activity is regulated by photosynthesis, is important in understanding how monospores are produced and discharged.

MIGRATION OF ASEXUAL MONOSPORES: ESTABLISHMENT OF CELL POLARITY

In the asexual life cycle of *P. yezoensis*, monospores are released from monosporangia and migrate while undergoing morphological change. They then adhere to the substratum and divide asymmetrically to produce two different vegetative and rhizoid cells (Guiry, 1990; Pickett-Heaps et al., 2001; Ackland et al., 2007; Li et al., 2008a,b; Mikami, 2010). Here, we summarize the current progress of studies aimed at understanding monospore development in *P. yezoensis*, focusing on the establishment of cell polarity during monospore migration (Li et al., 2008a,b; Li et al., 2009; Mikami et al., 2009, Mikami, 2010).

Involvement of the Cytoskeleton

During migration, F-actin became densely assembled at the leading edge (front side) of monospores (Figure 2Bb), although actin filaments were observed as bundles in undeveloped monospores (Figure 2Ba). Thus, migration of monospores requires pre-establishment of cell polarity for the asymmetrical localization of F-actin at the front side of migrating cells. In addition, when freshly released monospores were incubated with Cytochalasin B (digestion of actin filaments), Latrunculin B (inhibitor of G-actin polymerization), Nocodazole (inhibitor of MTs) or 2,3-butanedione monoxime (myosin inhibitor A), they did not move and maintained a spherical shape (Li et al., 2008a), indicating that the cytoskeleton organization plays a critical role in the early development of monospores. Therefore, establishment of an anterior-posterior axis is necessary for promoting the asymmetrical distribution of F-actin and other cytoskeleton elements, which produces the force for migration at the leading edge and organization of the cytoskeleton (Mikami, 2010).

Involvement of Photosynthetic Activity

Migration and development of monospores is completely inhibited in the dark (Li et al., 2009), indicating the importance of light illumination in monospore development. The requirement of light for polarity establishment was previously observed in Fucoid zygotes (Kropf, 1992; Brownlee et al., 2001). In these zygotes, the direction of light influenced the establishment of the cell axis as rhizoids grew away from light; however, sperm entry triggers the formation of a default axis in the dark with an F-actin patch, adhesive secretion and rhizoid outgrowth found at the position of sperm entrance, although the default axis is overridden by unilateral light (Henderson et al., 1998; Hable and Kropf, 2000).

In contrast, when monospores were irradiated with unilateral light, the direction of migration and light were not correlated (Li et al., 2009), indicating that migration and early development of germlings are not dependent on the direction of light. Thus, the regulatory mechanism involved in the establishment of cell polarity in monospores is different from that in Fucoid zygotes. In fact, treatment of monospores with DCMU strongly inhibited migration (Li et al., 2009), indicating that photosynthesis is required for the establishment of cell polarity via photosynthetic activity, which is required for migration (Mikami et al., 2009).

Involvement of Extracellular Ca^{2+} Influx

It is well known that spatio-temporal increases in cytoplasmic free Ca^{2+} ($[Ca^{2+}]_{cyt}$) are mainly generated by an influx of extracellular Ca^{2+} through membrane Ca^{2+}-permeable channels, allowing regulation of a variety of Ca^{2+}-dependent signaling systems in plants (Berridge et al., 2000; Sanders et al., 2002; Wheeler and Brownlee, 2008). It has also been reported that calcium gradients and calcium-dependent proteins are spatially and temporally regulated in tip-growing cells (Hepler et al., 2001; Yoon et al., 2006). Indeed, a tip-high $[Ca^{2+}]_{cyt}$ gradient is required for the polarized growth of pollen tubes and root hairs in higher plants and rhizoids in Fucoid brown algae (Homblé and Léonetti, 2007; Miedema et al., 2008; Chen et al., 2009). Moreover, a tip-polarized increase in $[Ca^{2+}]_{cyt}$, which is controlled by the direction of light, regulates the formation of the apical-basal axis during rhizoid development in Fucoid zygotes (Brownlee and Pulsford, 1998; Taylor et al., 1996). Thus, Ca^{2+} influx is fundamental in the formation of cell polarity.

In *P. yezoensis*, extracellular Ca^{2+} influx was also shown to be important for the establishment of cell polarity. Indeed, treatment of monospores with 1 mM EGTA or 100 μM $LaCl_3$, a Ca^{2+} channel blocker, prevented the asymmetrical distribution of F-actin and migration (Li et al., 2009). Moreover, the calcium ionophore A23187 counteracted the inhibition of migration in the dark (Li et al., 2009), indicating that photosynthetic activity regulates the Ca^{2+} influx to promote the establishment of cell polarity in monospores.

Taken together with the requirement of light illumination, it appears that a photosynthesis-mediated Ca^{2+} influx triggers the initiation of monospore migration. Although little is known about how photosynthesis regulates Ca^{2+} influx, the dependency of extracellular Ca^{2+} influx on photosynthesis is a novel system that increases cytosolic Ca^{2+} concentration in plants. Similar regulation has been reported in blue light-stimulated Ca^{2+} uptake via Ca^{2+} channels in guard cells (Harada and Shimazaki, 2009).

Critical Involvement of Phospholipase C in Cell Polarity Establishment

A rise in $[Ca^{2+}]_{cyt}$ is sensed by proteins such as Ca^{2+}-dependent kinases and phospholipase C (PLC), which subsequently activate downstream phosphoinositide signaling cascades (Saimi and Kung, 2002; Yoon et al., 2006; Kim et al., 2007). PLC hydrolyzes phosphatidylinositol-4,5-bisphosphate [PtdIns(4,5)P_2] in a Ca^{2+}-dependent manner via the

Ca^{2+}-binding EF hand motif (Kouchi et al., 2005), producing two second messengers, diacylglycerol (DG) and inositol 1,4,5-trisphosphate (IP3), which, in turn, activate protein kinase C (PKC) and Ca^{2+} release from intracellular stores via the IP3 receptor (IP3R), respectively (Berridge and Irvine, 1984). The involvement of PLC in polarity establishment in monospores has been confirmed by the treatment of freshly released monospores with 1µM U73122, a specific inhibitor of PLC (Smith et al., 1990), which resulted in the inhibition of movement and asymmetrical F-actin distribution (Li et al., 2009), although concentrations were much lower than those used in studies of higher plants and animals (10-100 µM). PLC is, therefore, involved in the establishment of cell polarity by directing the asymmetrical localization of F-action and subsequent migration of monospores (Li et al., 2009; Mikami et al., 2009).

PLC hydrolyzes PtdIns(4,5)P_2 into two second messengers, IP3 and DG (Katan, 1998). Since IP3R was recently shown to exist in green algae (Wheeler and Brownlee, 2008), the requirement of IP3R-like activity in the early development of monospores was examined using an IP3R antagonist, 2-APB, which inhibits IP3R activity on the ER membrane in animal cells (Maruyama et al., 1997). Treatment of monospores with 20 µM 2-APB prevented the migration of monospores, but caused erratic movement (Li et al., 2009). The other second messenger, phosphatidic acid (PA), is produced from DG through DGK in plants (Meijer and Munnik, 2003). Since PA is the main second messenger in plants, its role in the polarity formation of monospores was tested using the DGK inhibitor R59022. Treatment of freshly released monospores with 15 µM R59022 for three hours inhibited both the migration and asymmetric distribution of F-actin (Li et al., 2009).

The disruption of F-actin asymmetry by both 2-APB and R59022 suggest that IP3R-like protein and DGK are involved in the establishment of cell polarity in monospores (Li et al., 2009; Mikami et al., 2009). However, important questions about the presence of IP3R in *P. yezoensis* cells remain.

Involvement of PI3K in Migration

The molecular mechanisms regulating the asymmetric distribution of F-actin have been studied extensively in *Dictyostelium* cells and leukocytes. One of the first asymmetric responses to chemoattractants is the localized accumulation of phosphatidylinositol-3,4,5-trisphosphate [PtdIns(3,4,5)P_3], the product of PI3K (Merlot and Firtel, 2003; Dormann et al., 2004). In the new leading edge of *Dictyostelium* cells, preferential activation of PI3K at the side facing the chemoattractant gradient is necessary for polarized F-actin localization and directional movement (Parent and Devreotes, 1999; Firtel and Chung, 2000; Bourne and Weiner, 2002). In contrast, phosphatase and Tensin homolog (PTEN), which dephosphorylates PtdIns(3,4,5)P_3, localizes on the trailing edge where it acts as a negative regulator of PI3K signaling in *Dictyostelium* cells (Iijima et al., 2002). In addition, Nishio et al. (2007) argued that Src homology 2 domain-containing inositol-5-phosphatase 1 (SHIP1), not PTEN, is the key regulator of neutrophil migration. The localized distribution of PI3K and PtdIns(3,4,5)P_3 phosphatase, therefore, helps cells define their polarity by organizing the polarized localization of F-actin (Charest and Firtel, 2006).

Migrating monospores form a tapered shape as observed in *Dictyostelium* cells and leukocytes (Figure 3B; Li et al., 2008a; Mikami, 2010), suggesting the possible involvement of PI3K and PtdIns(3,4,5)P_3-phosphatase in the movement of *P. yezoensis* monospores. Indeed, polarized localization of F-actin was prevented after three hours of incubation with LY294002, the specific inhibitor of PI3K (Li et al., 2008a), indicating that PI3K activity is involved in migration through formation of the anterior-posterior axis and the regulation of asymmetrical F-actin accumulation at leading edges (Li et al., 2008a; Mikami, 2010).

It has been established that PTEN and SHIP1 are negative regulators of PtdIns(3,4,5)P_3 production by PI3K in *Dictyostelium* cells and leukocytes (Iijima et al., 2002; Manahan et al., 2004; Nishio et al., 2007; Oudit and Penninger, 2009). Although the possible existence of SHIP1-like activity in monospores could not be examined, PTEN-like activity was tested using bpv(pic), a PTEN-specific inhibitor (Schmid et al., 2004). As a result, it was concluded that PTEN-like activity does not function during directional migration of monospores (Li et al., 2008a).

Maintenance of Cell Polarity During Migration of Monospores

Involvement of PLD

DG produced by PLC is immediately phosphorylated by DGK to produce PA (Munnik, 2001; Meijer and Munnik, 2003). PA activates phosphatidylinositol phosphatase kinase (PIPK) to produce PtdIns(4,5)P2 as a substrate of PLC in both animals and plants (Oude Weernink et al., 2007; Saavedra et al., 2009). PtdIns(4,5)P2 then activates PLD, which hydrolyzes phosphatidylcholine to produce PA (Moritz et al., 1992; Jenkins et al., 1994; Ishihara et al., 1998; Jones et al., 2000). Thus, it appears that both PLC and PLD exert control over PA concentrations in plant cells. However, there are differences in the roles of PLC and PLD in the formation of cell polarity. For example, PLC regulates F-actin dynamics, vesicle trafficking and ion transport in pollen tubes (Hunt et al., 2003; Dowd et al., 2006; Helling et al., 2006), whereas PLD regulates the organization of microtubules in Fucoid embryos as well as seed germination and root elongation in green plants (Gardiner et al., 2001, 2003; Dhonukshe et al., 2003; Peters et al., 2007)..

Recently, the roles of PLD in cell polarity establishment have been covered in red algae. Treatment of monospores with 0.4% 1-butanol, a PLD inhibitor, for three hours had no effect on asymmetrical F-actin localization; however, eight-hour treatment of monospores resulted in symmetrically distributed F-actin. Thus, PLD participates in the maintenance, but not establishment, of cell polarity during the early development of monospores (Li et al., 2009; Mikami et al., 2009).

Involvement of Cell Wall Synthesis

The extensive knowledge on the cell wall of Fucoid zygotes indicates their involvement in the maintenance of cell polarity (Brownlee and Bouget, 1998; Bisgrove and Kropf, 2001a). It is well known that the cell wall plays an essential role in the establishment of cell polarity, a phenomenon documented in the brown algae *Fucus* (Kropf et al., 1988; Fowler and

Quatrano, 1997; Belanger and Quatrano, 2000). Zygotes of *Fucus* have the ability to establish an axis in accordance with an orienting vector of unilateral light (Murphy et al., 1970). Moreover, cell wall synthesis is an absolute requirement for axis fixation because the orientation of the axis remains labile until a new cell wall is synthesized (Kropf et al., 1988), for which polarized secretion via Golgi apparatus is needed (Belanger and Quatrano, 2000).

Although freshly released monospores are cell wall free as zygotes of Fucoid algae, synthesis of the cell wall is initiated during cell migration (Li et al., 2008b). When freshly released monospores were treated for three hours with brefeldin A (BFA), a specific inhibitor of polysaccharide biosynthesis required for cell wall formation via Golgi-derived vesicle trafficking, F-actin normally accumulated at the leading edge in tapered-shape migrating monospores, whereas the polarized localization of F-actin was disrupted after 20 hours of BFA treatment (Li et al., 2008b). Thus, the cell wall is not required for the establishment of the anterior-posterior axis, but is indispensable for the maintenance of cell polarity and the asymmetrical distribution of F-actin during monospore migration.

Adherence of Monospores: Axis Conversion for Gametophyte Development

Development of leafy gametophytes is performed along with the apical-basal axis, whose establishment is due to the conversion of the anterior-posterior axis by cessation in the migration and subsequent adhesion of monospores to the substratum (Figure 2Bc). In *P. yezoensis*, although synthesis of the cell wall is observed during migration in monospores, BFA treatment of monospores resulted in the inhibition of both synthesis of calcofluor-staining materials and adhesion to the substratum (Li et al., 2008b). Thus, Golgi-mediated vesicle transport is thought to be required for the secretion of adhesion materials and subsequent axis conversion. However, little is known about how polarized intracellular vesicle translocation and axis conversion is regulated in *P. yezoensis*.

Determination of Cell Fate via Asymmetric Cell Division

Symmetry breaking is universal machinery to produce asymmetry for choosing differentiation of cells, since the asymmetrical distribution of cellular components leads to the production of daughter cells with different fates after cell division (tenHove and Heidstra, 2008; Menke and Scheres, 2009). For example, during embryogenesis of angiosperms, the zygote elongates along with the apical-basal axis, which has already been established in female gametophytes to allow the asymmetric localization of the nucleus and vacuole (Mansfield and Briatry, 1991; de Jong et al., 1993). It then divides asymmetrically to produce a smaller apical cell and a large basal cell; further cell division of the former leads to the development of cotyledon tissue still carrying the apical-basal axis, whereas that of the latter results in the development of suspensor cells (tenHove and Heidstra, 2008; De Smet et al., 2010).

Involvement of the apical-basal axis in the establishment of intracellular asymmetry is well known in zygotic development of brown fucoid algae (Brownleee and Bouget, 1998;

Bisgrove, 2007). Unidirectional light induces rhizoid formation on the shaded hemisphere via F-actin accumulation and localized secretion of membrane, cell wall and adhesion materials at the rhizoid pole, which brings about morphological change in the spherical zygote causing a pear-shaped cell with apical-basal axis determining thallus-rhizoid poles (Hable et al., 2003; Hable and Kropf, 2005; Hadley et al., 2006). Subsequently, transverse and asymmetric cell division occurs, generating two morphologically distinct cells: a thallus cell that develops into a gametophyte and a rhizoid cell that becomes the holdblast (Bisgrove, 2007), in which the position of the division plane is determined by the centrosome-dependent microtubule array (Bisgrove and Kropf, 2001b).

During the development of monospores, conversion of the cell axis is subjected to first cell division occurring perpendicularly to the apical-basal axis (Figure 2Be). It is worth noting that the upright form of monospores elongates along with the apical-basal axis after adhesion to the substratum (Figure 2Bd), which is similar to angiosperm zygotes that elongate to produce asymmetry before cell division (Mansfield and Briatry, 1991; de Jong et al., 1993). In *P. yezoensis*, since F-actin accumulates at the rhizoid pole in adhering and elongating *P. yezoensis* monospores (Figure 2Bc,d), it is possible that F-actin plays a role in determination of the rhizoid pole along with the apical-basal axis during monospore development. In addition, it is also characteristic that the first asymmetric cell division in monospores divides the chloroplast into two symmetric parts without fail (Figure 2Be). Thus, positioning of the mitotic apparatus could be determined by a mechanism that recognizes the position of the chloroplast. However, little is known about how cellular differentiation via asymmetrical cell division is regulated in monospores. In angiosperms, the genes encoding regulators of intracellular asymmetry, asymmetric cell division and molecular determinants of cell fates have been identified by mutational and reverse-genetic approaches (tenHove and Heidstra, 2008; Menke and Scheres, 2009; De Smet et al., 2010).

Comparison of Asexual Spore Production between *P. yezoensis* and Filamentous Fungi

A spore-based asexual life cycle is also found in filamentous fungi such as *Aspergillus nidulans* and *Neurospora crassa* (Springer, 1993; Adams et al., 1998; Osherov and May, 2001). Interestingly, there is a fundamental difference in asexual spore formation between *P. yezoensis* and filamentous fungi. In filamentous fungi, asexual spores, known as conidia, are produced in specially differentiated conidiophores. Briefly, a filamentous branch elongates by apical extension to produce a conidiophore stalk, the tip region of which swells producing a conidiophore vesicle. Nuclear division then occurs at the surface of the conidiophore vesicle, resulting in the production of ca 60 cigar-shaped cells (matulaes), each of which contains a single nucleus. Next, the number of matulaes doubles giving ca 120 uninucleated phialides, from which a chain of uninuclear spores (conidia) is produced via repeated asymmetric cell division of the phialides themselves. As a result, ca 1,000 conidia are usually made from a single phialide. Thus, phialides are thought to be stem cells.

In contrast, multi-step cellular differentiation is not observed in the production of asexual monospores in *P. yezoensis*. As mentioned above, initiation of monospore production is recognized by a change in vegetative cells to deep-red and cytoplasm-less monosporangia cells (Figure 2Ab). There are two possibilities as to how monospores are produced in monosporangia. One is differentiation via the transition of vegetative cells into spore cells without cell division in monosporangia. A representative example of the transition of a non-stem cell without cell division is the differentiation of heterocysts in filamentous cyanobacteria such as *Anabaena* sp. PCC7120 (*Nostoc* sp. PCC7120) and *Anabaena cylindrical*. In these species, it is well established that a signal triggering the initiation of heterocyst development from vegetative cells and signaling transduction pathways leading gene expression required for heterocyst development are involved (Flores and Herrero, 2010; Kumar et al., 2010). The other possibility, as is the case in filamentous fungi, is the asymmetric division of stem cells located at the boundary of monosporangia and vegetative cell region. Identification of the genes involved in the initiation of monospore production is necessary for resolution of these possibilities and understanding how and why *P. yezoensis* selects asexual propagation in addition to the sexual life cycle.

Similarities between Early Monospore Development and Epithelial-Mesenchymal Transition (EMT) in Vertebrates

As mentioned above, the monospore-based asexual life cycle consists of monosporangia formation, discharge of monospores from gametophytes, migration and adhesion, and development of monospores into gametophyte clones. In other words, this process can be recognized as vegetative cell-monospore-vegetative cell transition via de-polarization and re-polarization. In fact, a similar system has already been described in vertebrates, namely, epithelial-mesenchymal transition (EMT). In EMT, polarized epithelial cells are converted into mesenchymal cells with a loss of cell polarity followed by differentiation into specialized cell types during gastrulation and organ development (Kalluri and Weinberg, 2009; Thiery et al., 2009). Mesenchymal cells present cellular characteristics found in monospores, such as migration and cell adhesion capacities. In addition, a reverse process of EMT is also found, namely, mesenchymal to epithelial transition (MET; Thiery et al., 2009). Thus, it is possible that the monospore-based asexual life cycle in *P. yezoensis* is recognized as an ancient form of the EMT-MET process, although the regulatory systems of each are thought to differ significantly. Elucidation of the mechanisms regulating loss of cell polarity and its subsequent re-establishment during the production and development of monospores could provide novel and fundamental information about differentiation through cellular conversion in eukaryotes.

Biological Significance of the Asexual Life Cycle in Bangiophycideae

To date, it remains to be settled why some Bangiophycean red algae have retained both sexual and asexual life cycles and also the significance of the asexual life cycle compared with the sexual life cycle in such species needs to be determined. Due to the dominance of the sexual life cycle in red algae, we propose that the advantage of the asexual life cycle is related to the relatively low success rate of syngamy between male and female gametes. Red algal cells are aflagellate during their entire life cycle (Tripodi and de Masi, 1983; Clayton, 1992);

thus, success of syngamy depends completely on the number of freely discharged male gametes compared with the number of female gametes located in gametophytes. The probability of syngamy may be enhanced by increasing the number of free male gametes and the total area producing female gametes in gametophytes. Thus, an increase in the density of gametophytic clones produced by the asexual propagation of monospores could increase both the number of male gametes released in sea water and the probability of fertilizing female gametes existing in gametophytes.

The above suggestion is supported by the multiple patterns of sexual gamete formation observed in *Porphyra* species. As shown in Figure 4, we found a close relationship between monospore production and the pattern of sexual gamete production in gametophytes. The high production of monospores is usually observed in hermaphroditic species such as *P. yezoensis* and *P. tenera*, whose sexual gametes are produced in randomly mixed male and female regions in gametophytes. Monospore-less species are classified into two groups: one includes hermaphroditic species such as *P. variegata* and *P. purpurea*, in which gametophytes are segregated longitudinally into two regions for the production of male and female gametes; the other includes dioecious species such as *P. pseudolinearis*, *P. dentata*, and *P. haitanensis*, in which male and female gametophytes develop separately (Figure 4). Since the area for production of gametes is larger in monospore-less species than in monospore-producing species (Figure 4), it is possible to increase the chance of free aflagellate male gametes meeting female gametes located in the gametophytes. Thus, the production and release of male gametes in monospore-less species is comparable to the increase in gametophytic clones by the production of monospores. Alternatively, the asexual life cycle in monospore-producing species is important for an increased rate of syngamy against the small population of male gametes produced in a limited area of the gametophytes.

In fact, it has been reported that some species, such as *P. katadae* and *P. purpurea* showing the longitudinal segregation pattern of gamete production (Figure 4), often produce unisexual male gametophytes (Miura, 1968; Bray et al., 2006). This unique strategy also contributes to an increase in both the number of male gametes and successful syngamy in monospore-less *Porphyra* species.

A spore-mediated increase in the success rate of syngamy is also found in the triphasic life cycle in Florideophyceae, another class of Rhodophyte (Thornber, 2006). In this life cycle, a diploid zygote grows by attachment to a haploid female gametophyte as a carposporophyte, in which multiple diploid carpospores are then produced by mitosis. The carpospores are then released and develop into a diploid tetrasporophyte. When the tetrasporophyte matures, meiosis occurs to produce haploid tetraspores, which then grow into male and female gametophytes that separately produce male and female gametes for syngamy. Thus, the increase in the number of diploid carpospores compensates for the relatively infrequent occurrence of syngamy (Searles, 1980), which is, in fact, an identical explanation for the advantage of the asexual life cycle in *Porphyra* species given above. However, although the triphasic life cycle in Florideophyceae and the biphasic life cycle in Bangiophycideae share a strategy for increasing the rate of syngamy, the mechanisms regulating the production of spores completely differ between triphasic and biphasic life cycles in Rhodophyta.

Hermaphrodite		Dioecism
Sexual and asexual life cycle	Sexual life cycle	Sexual life cycle
Monospore production	No asexual spore	No asexual spore
Mosaically mixed pattern 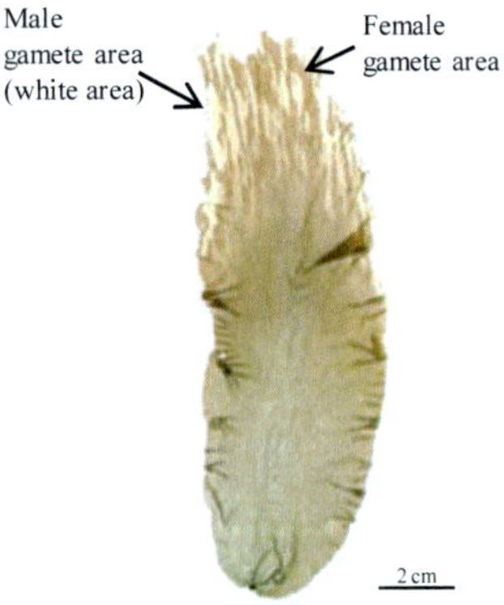	Longitudinally sectored pattern 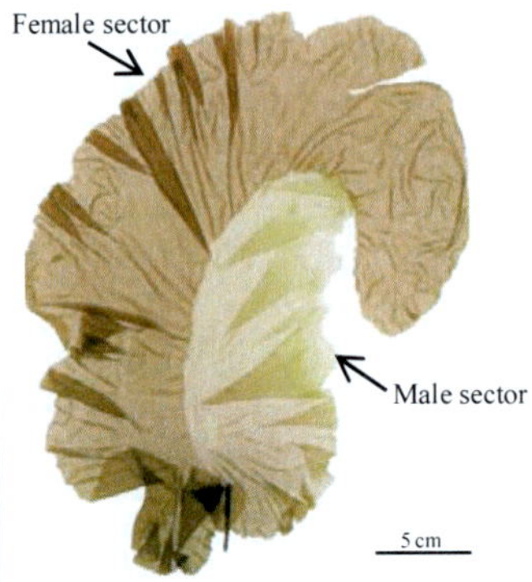	
P. yezoensis	*P. variegata*	*P. pseudolinearis*
P. tenera *P. okamurae* *P. onoi* *P. pulchella*	*P. katadae* *P. purpurea* There speices produce unisexual male gametophytes in addition to normal mature blades.	*P. dentata* *P. haitanensis* *P. angusta*

Figure 4. Three representative types of sexual patterning observed in Porphyra gametophytes According to the style of gamete production, *Porphyra* species are divided into three groups. Of these, monospores are produced only in hermaphrodite species whose sexual gametes are produced in randomly mixed male and female regions in the gametophyte. *P. yezoensis* belongs to this group. Monospore-less species having only the sexual life cycle are divided into two groups, hermaphrodite and dioecious species. The former shows segregated male and female areas in distinct longitudinal halves of the gametophyte, whereas the latter produces separate free-living male and female gametophytes.

CONCLUSION

The presence of asexual reproduction via monospores is a remarkable survival strategy in Bangiophycean red algae because of its absence in land plants. In *P. yezoensis*, both the production/discharge and migration of asexual monospores with the establishment of an anterior-posterior axis require a photosynthesis-dependent extracellular Ca^{2+} influx. Based on these findings, it is hypothesized that Ca^{2+} influx mediated by probable Ca^{2+} permeable channels and a resultant increase in $[Ca^{2+}]cyt$ are indispensable for the establishment of cell polarity in monospores and subsequent migration and development. Therefore, identification of the genes encoding these channels and targets of Ca^{2+} is necessary for further molecular biological and genetic studies on the regulation of the asexual life cycle in eukaryotic red algae.

The biological significance of the asexual life cycle in *P. yezoensis* is thought to be related to an increase in the frequency of successful syngamy based on the aflagellate nature of male gametes. Since syngamy in the ancestors of Bangiophycean red algae appears to have been inefficient, it is hypothesized that the monospore-based asexual life cycle was evolutionally established to improve the success rate of syngamy by producing a large number of monospores from gametophytic clones. Elucidation of the molecular basis of the monospore-based asexual life cycle could enable understanding of how and why the asexual life cycle was established as a unique survival strategy in Bangiophycideae.

Acknowledgments

We are grateful to Dr. Chikara Kawagoe (Algatech Kyowa, Kyowa Concrete Industry Co. Ltd.) and Ryo Hirata (Hokkaido University) for assistance with collection of naturally growing *P. yezoensis*, *P. variegata* and *P. pseudolinearis*. This work was supported by a grant from the Hokusui Society Foundation.

References

Ackland, J. C., West, J. A. and Pickett-Heaps, J. (2007). Actin and myosin regulate pseudopodia of *Porphyra pulchella* (Rhodophyta) archeospores. *Journal of Phycology, 43,* 129-138.

Adams, T. H., Wieser, J. K. and Yu, J. H. (1998). Asexual sporulation in *Aspergillus nidulans*. *Microbiology and Mollecular Biology Reviews, 62,* 35-54.

Batters, E. A. L. (1892). On *Conchocelis,* a new genus of perforating algae. In G. Murray (Ed.), *Phycological Memoirs* (pp. 25-27). London. Dulau.

Belanger, K. D. and Quatrano, R. S. (2000). Polarity: the role of localized secretion. Current *Opinion in* Plant Biology, 3, *67-72.*

Berridge, M. J. and Irvine, R. F. (1984). Inositol trisphosphate, a novel second messenger in cellular signal transduction. *Nature, 312,* 315-321.

Berridge, M. J., Lipp, P. and Bootman, M. D. (2000). The versatility and universality of calcium signaling. *Nature Reviews Molecular Cell Biology, 1,* 11-21.

Bisgrove, S. R. (2007). Cytoskeleton and early development in fucoid algae. *Journal of Integrative Plant Biology, 49,* 1192-1198.

Bisgrove, S. R. and Kropf, D. L. (2001a). Cell wall deposition during morphogenesis in fucoid algae. *Planta, 212,* 648-58.

Bisgrove, S. R. and Kropf, D. L. (2001b). Asymmetric cell division in fucoid algae: a role for cortical adhesion in alignment of the mitotic apparatus. *Journal of Cell Science*, 114, 4319-4328.

Bourne, H. R. and Weiner, O. (2002). A chemical compass. *Nature, 419,* 21.

Bray, T.L., Neefus, C.D. and Mathieson, A.C. (2006). Morphological and molecular variability of *Porphyra purprea* (Roth) C. Agardh (Rhodophyta, Bangiales) from the Northwest Atlantic. *Nova Hedwigia, 82,* 1-22.

Brownlee, C. and Bouget, F. Y. (1998). Polarity determination in *Fucus*: from zygote to multicellular embryo. Seminars *in* Cell *and* Developmental Biology, *9,* 179-185.

Brownlee, C., Bouget, F. Y. and Corellou, F. (2001). Choosing sides: establishment of polarity in zygotes of fucoid algae. *Seminars in Cell and Developmental Biology, 12,* 345-351.

Butterfield, N. J. (2000). *Bangiomorpha pubescens* n. gen., n. sp.: implications for the evolution of sex, multicellularity, and the Mesoproterozoic/ Neoproterozoic radiation of eukaryotes. *Paleobiology, 26,* 386-404.

Charest, P. G. and Firtel, R. A. (2006). Feedback signaling controls leading-edge formation during chemotaxis. *Current Opinion in Genetics and Development, 16,* 339-347.

Charnofsky, K., Towill, L. R., and Sommerfeld, M. R. (1982). Light requirements for monospore germination in *Bangia atropurpurea* (Rhodophyta). *Journal of Phycology, 18,* 417-422.

Chen, T., Wu, X., Chen, Y., Li, X., Huang, M., Zheng, M., Baluska, F., Samaj, J. and Lin, J. (2009). Combined proteomic and cytological analysis of Ca^{2+}-calmodulin regulation in *Picea meyeri* pollen tube growth. *Plant Physiology, 149,* 1111-1126.

Clayton, M. N. (1992). Propagules of marine macroalgae: Structure and development. *European Journal of Phycology, 27,* 219-232.

Coelho, S. M., Peters, A. F., Charrier, B., Roze, D., Destombe, C., Valero, M. and Cock, J. M. (2007). *Complex life cycles of multicellular eukaryotes*: *new approaches based* on the use of model organisms. *Gene, 406,* 152-170.

de Jong, A.J., Schmidt, E.D.L. and de Vries, S.C. (1993) Early events in higher-plant embryogenesis. *Plant Molecular Biology*, 22, 369-377.

De Smet, I., Lau, S, Mayer, U. and Jürgens, G. (2010) Embryogenesis–the humble beginnings of plant life. *Plant Journal*, 61, 959-970.

Dhonukshe, P., Laxalt, A. M., Goedhart, J., Gadella, T. W. J. and Munnik, T. (2003). Phospholipase D activation correlates with microtubule reorganization in living plant cells. *Plant Cell, 15,* 2666-2679.

Dormann, D., Weijer, G., Dowler, S. and Weijer, C. J. (2004). *In vivo* analysis of 3-phosphoinositide dynamics during *Dictyostelium* phagocytosis and chemotaxis. *Journal of Cell Science, 117,* 6497-6509.

Dowd, P. E., Coursol, S., Skirpan, A. L., Kao, T. H. and Gilroy, S. (2006). *Petunia* phospholipase C1 is involved in pollen tube growth. *Plant Cell, 18,* 1438-1453.

Drew, K. M. (1949). *Conchocelis-phase in the life-history of* Porphyra umbilicalis (L.) Kütz. *Nature, 166,* 748-749.

Firtel, R. A. and Chung, C. Y. (2000). The molecular genetics of chemotaxis: sensing and responding to chemoattractant gradients. *BioEssays, 22*, 603-615.

Flores, E. and Herrero, A. (2010). Compartmentalized function through cell differentiation in filamentous cyanobacteria. *Nature Reviews Microbiology, 8,* 39-50.

Fowler, J. E. and Quatrano, R. S. (1997). Plant cell morphogenesis: plasma membrane interactions with the cytoskeleton and cell wall. Annual *Review of* Cell *and Developmental* Biology, *13,* 697-743.

Gardiner, J., Collings, D. A., Harper, J. D. I. and Marc, J. (2003). The effects of the phospholipase D-antagonist 1-butanol on seedling development and microtubule organization in *Arabidopsis*. *Plant cell Physiology, 44,* 687-696.

Gardiner, J. C., Harper, J. D. I., Weerakoon, N. D., Collings, D. A., Ritchie, S., Gilroy, S., Cyr, R. J. and Marc, J. (2001). A 90-kD phospholipase D from tobacco binds to microtubules and the plasma membrane. *Plant Cell, 13,* 2143-2158.

Guiry, M. D. (1990). The life history of *Liagora harveyana* (Nemaliales, Rhodophyta) from south-eastern Australia. *European Journal of Phycology, 25,* 353-362.

Hable, W. E. and Kropf, D. L. (2000). Sperm entry induces polarity in focoid zygotes. *Development, 127,* 493-501.

Hable, W. E. and Kropf, D. L. (2005). The Arp2/3 complex nucleates actine arrays during xygote polarity establishment and growth. *Cell Motility and Cytoskeleton, 61,* 9-20.

Hable, W. E., Miller, N.R. and Kropf, D. L. (2003). Polarity establishment requires dynamic actin in fucoid zygotes. *Protoplasma, 221*, 193-204.

Hadley, R., Habel, W.E. and Kropf, D.L. (2006). Polarization of the endomembrane system is an early event in fucoid zygote development. *BMC Plant Biology, 6,* 5 (doi:10.1186/1471-2229-6-5).

Harada, A. and Shimazaki, K. (2009). Measurement of changes in cytosolic Ca^{2+} in *Arabidopsis* guard cells and mesophyll cells in response to blue light. *Plant and Cell Physiology, 50,* 360-373.

Helling, D., Possart, A., Cottier, S., Klahre, U. and Kost, B. (2006). Pollen tube tip growth depends on plasma membrane polarization mediated by tobacco PLC3 activity and endocytic membrane recycling. *Plant Cell, 18,* 3519-3534.

Henderson, D. C., Bisgrove, S. R., Hable, W. E., Alessa, L. and Kropf, D. L. (1998). Division patterns in the thallus of *Pelvetia compressa* embryos and the effects of gravity. *Protoplasma, 203,* 112-117.

Hepler, P. K., Vidali, L. and Cheung, A. Y. (2001). Polarized cell growth in higher plants. *Annual Review of Cell and Developmental Biology, 17,* 159-187.

Homblé, F. and Léonetti, M. (2007). Emergence of symmetry breaking in fucoid zygotes. *Trends in Plant Science, 12,* 253-259.

Hunt, L., Mills, L. N., Pical, C., Leckie, C. P., Aitken, F. L., Kopka, J., Mueller-Roeber, B., McAinsh, M. R., Hetherington, A. M. and Gray, J. E. (2003). Phospholipase C is required for the control of stomatal aperture by ABA. *Plant Journal, 34,* 47-55.

Iijima, M., Huang, Y. E. and Devreotes, P. (2002). Temporal and spatial regulation of chemotaxis. *Developmental Cell, 3,* 469-478.

Ishihara, H., Shibasaki, Y., Kizuki, N., Wada, T., Yazaki, Y., Asano, T, and Oka, Y. (1998). Type I phosphatidylinositol-4-phosphate 5-kinases. Cloning of the third isoform and deletion/substitution analysis of members of this novel lipid kinase family. *Journal of Biological Chemistry, 273,* 8741-8748.

Jenkins, G. H, Fisette, P. L. and Anderson, R. A. (1994). Type I phosphatidylinositol 4-phosphate 5-kinase isoforms are specifically stimulated by phosphatidic acid. *Journal of Biological Chemistry, 269,* 11547-11554.

Jones, D. R., Sanjuan, M. A. and Mérida, I. (2000). Type Iα phosphatidylinositol 4-phosphate 5-kinase is a putative target for increased intracellular phosphatidic acid. *FEBS Letter, 476,* 160-165.

Kalluri, R. and Weinberg, R. A. (2009). The basics of epithelial-mesenchymal transition. *Journal of Clinical Investigation, 119*, 1420-1428.

Katan, M. (1998). Families of phosphoinositide-specific phospholipase C: structure and function. *Biochimica et Biophysica Acta, 1436,* 5-17.

Kim, B. G., Waadt, R., Cheong, Y. H., Pandey, G. K., Dominguez-Solis, J. R., Schültke, S., Lee, S. C., Kudla, J. and Luan, S. (2007). The calcium sensor CBL10 mediates salt tolerance by regulating ion homeostasis in *Arabidopsis*. *Plant Journal, 52,* 473-484.

Kitade, Y., Fukuda, S., Nakajima, M., Watanabe, T. and Saga, N. (2002). Isolation of a cDNA encoding a homolog of actin from *Porphyra yezoensis* (Rhodophyta), *Journal Applied of Phycology, 14,* 135–141.

Kitade, Y., Taguchi, G., Shin, J.-A. and Saga, N. (1998). *Porphyra* monospore system (Bangiales, Rhodophyta): a model for the developmental biology of marine plans. *Phycological Research, 46,* 17-20.

Kito, H. (1974). Cytological observations on the conchocelis–phase in three species of *Porphyra* (in Japanese with English summary). Bulletin *of* Tohoku Regional Fisheries Research Laboratory, *33,* 101-117.

Kouchi, Z., Shikano, T., Nakamura, Y., Shirakawa, H., Fukami, K. and Miyazaki, S. (2005). The role of EF-hand domains and C2 domain in regulation of enzymatic activity of phospholipase Cζ. *Journal of Biological Chemistry, 280,* 21015-21021.

Kropf, D. L. (1992). Establishment and expression of cellular polarity in Fucoid zygotes. *Microbiological Reviews, 56,* 316-339.

Kropf, D. L., Kloareg, B. and Quatrano, R. S. (1988). Cell wall is required for fixation of the embryonic axis of *Fucus* zygotes. *Science, 239,* 187-190.

Kumar, K., Mella-Herrera, R. A. and Golden, J. W. (2010). Cyanobacterial heterocysts. *Cold Spring Harbor Perspectives in Biology,* 10.1101/cshperspect.a000315.

Kurogi, M. (*1953).* Study of the *life-history* of Porphyra I. The germination and development of carpospores. Bulletin *of* Tohoku Regional Fisheries Research Laboratory*, 2,* 67-103. (in Japanese with English abstract).

Kurogi, M. (1961). Species of cultivated *Porphyra* and their life histories (Study of the life history of *Porphyra* II). *Bulletin of Tohoku Regional Fisheries Research Laboratory*, *18,* 1-115.

Li , L., Saga, N. and Mikami, K. (2008a). Phosphatidylinositol 3-kinase activity and asymmetrical accumulation of F-actin are necessary for establishment of cell polarity in the early development of monospores from the marine red alga *Porphyra yezoensis*. *Journal of Experimental Botany, 59,* 3575-3586.

Li , L., Saga, N. and Mikami, K. (2008b). Effects of cell wall synthesis on cell polarity in the red alga *Porphyra yezoensis. Plant Signaling and Behavior, 3,* 1126-1128.

Li , L., Saga, N. and Mikami, K. (2009). Ca^{2+} influx and phosphoinositide signaling are essential for the establishment and maintenance of cell polarity in monospores from the red alga *Porphyra yezoensis*. *Journal of Experimental Botany, 60,* 3477-3489.

Li, S. Y. (1984). The ecological characteristics of monospores of *Porphyra yezoensis* Ueda and their use in cultivation. *Hydrobiologia, 116,* 255-258.

Ma, J. H. and Miura, A. (1984). Observations of the nuclear division in the conchospores and their germlings in *Porphyra yezoensis* Ueda (in Japanese with English summary). *Japanese Journal of Phycology*, *32,* 373–378.

Mable, B. K. and Otto, S. P. (1998). The evolution of life cycles with haploid and diploid phases. *BioEssays*, *20*, 453-462.

Magne, F. (1991). Classification and phylogeny in the lower Rhodophyta: a new proposal. *Journal of Phycology, 27,* 46.

Manahan, C. L., Iglesias, P. A., Long, Y. and Devreotes, P. N. (2004). Chemoattractant signaling in *Dictyostelium discoideum*. *Annual Review of Cell and Developmental Biology, 20,* 223-253.

Mansfield, S. G. and Briarty, L. G. (1991) Early embryogenesis in *Arabidopsis thaliana*. II. The developing embryo. *Canadian Journal of Botany*, *69*, 461-476.

Maruyama, T., Kanaji, T., Nakade, S., Kanno, T. and Mikoshiba, K. (1997). 2APB, 2-aminoethoxydiphenyl borate, a membrane-permeable modulator of Ins (1,4,5)P3-induced Ca^{2+} release. *The Journal of Biochemistry, 122,* 498-505.

Meijer, H. J. G. and Munnik, T. (2003). Phospholipid-based signaling in plants. *Annual Review of Plant Biology, 54,* 265-306.

Menke, F. L. H. and Scheres, B. (2009) Plant asymmetric cell division, vive la différence! *Cell, 137,* 1189-1192.

Merlot, S. and Firtel, R. A. (2003). Leading the way: directional sensing through phosphatidylinositol 3-kinase and other signaling pathways. *Journal of Cell Science, 116,* 3471-3478.

Miedema, H., Demidchik, V., Véry, A. A., Bothwell, J. H., Brownlee, C. and Davies, J. M. (2008). Two voltage-dependent calcium channels co-exist in the apical plasma membrane of *Arabidopsis thaliana* root hairs. *New Phytologist, 179,* 378-385.

Migita, S. (1967). Cytological studies on *Porphyra yezoensis* Ueda. *Bulletin of the Faculty of Fisheries, Nagasaki University, 24,* 55–64.

Mikami, K., Li, L., Takahashi, M. and Saga, N. (2009). Photosynthesis-dependent Ca^{2+} influx and functional diversity between phospholipases in the formation of cell polarity in red algal migrating cells. *Plant Signaling and Behavior, 4,* 911-913.

Mikami, K. (2010). Migrating plant cell: F-actin asymmetry directed by phosphoinositide signaling. In S. Lansing and T. Rousseau (Eds.), *Cytoskeleton: Cell Movement, Cytokinesis and Organelles Organization* (pp. 205-218). New York, Nova Science Publishers.

Miura, A. (1968). *Porphyra katadai*, a new species from Japanese coast. *Journal of the Tokyo University of Fisheries, 54,* 55-59.

Miura, A. (1985). Genetic analysis of the variant color types of light red, light green and light yellow phenotypes of *Porphyra yezoensis* (Rhodophyta, Bangiaceae). In H. Hara (Ed.), *Origin and Evolution of Diversity in Plants and Plant Communities* (pp. 270-284). Tokyo, Academia Scientific Book Inc.

Moritz, A., De Graan, P. N., Gispen, W. H. and Wirtz, K. W. (1992). Phosphatidic acid is a specific activator of phosphatidylinositol-4-phosphate kinase. *The Journal of Biological Chemistry, 267,* 7207-7210.

Munnik, T. (2001). Phosphatidic acid: an emerging plant lipid second messenger. *Trends in Plant Science, 6,* 227-233.

Murphy, T. M., Kahn, A. and Lang, A. (1970). Stabilization of the polarity axis in the zygotes of some *Fucaceae*. *Planta, 90,* 97-108.

Nishio, M., Watanabe, K., Sasaki, J., Taya, C., Takasuga, S., Iizuka, R., Balla, T., Yamazaki, M., Watanabe, H., Itoh, R., Kuroda, S., Horie, Y., Förster, I., Mak, T. W., Yonekawa, H., Penninger, J. M., Kanaho, Y., Suzuki, A. and Sasaki, T. (2007). Control of cell polarity and motility by the PtdIns(3,4,5)P3 phosphatase SHIP1. *Nature Cell Biology, 9,* 36-44.

Ohme, M. and Miura, A. (1988). Tetrad analysis in conchospore germlings of *Porphyra yezoensis* (Rhodophyta, Bangiales). *Plant Science, 57,* 135–140.

Ohme, M., Kunifuji, Y. and Miura, A. (1986) Cross experiments of the color mutants in *Porphyra yezoensis* Ueda. *Japanese Journal of Phycology, 34,* 101–106.

Osherov, N. and May, G. S. (2001) The molecular mechanisms of conidial germination. *FEMS Microbiology Letters, 199,* 153-160.

Oude Weernink, P. A., Han, L., Jakobs, K. H. and Schmidt, M. (2007). Dynamic phospholipid signaling by G protein-coupled receptors. *Biochimica et Biophysica Acta, 1768,* 888-900.

Oudit, G. Y. and Penninger, J. M. (2009). Cardiac regulation by phosphoinositide 3-kinases and PTEN. *Cardiovascular Research, 82,* 250-260.

Parent, C. A. and Devreotes, P. N. (1999). A cell's sense of direction. *Science, 284,* 765-770.

Peters, N. T., Logan, K. O., Miller, A. C. and Kropf, D. L. (2007). Phospholipase D signaling regulates microtubule organization in the fucoid alga *Silvetia compressa. Plant and Cell Physiology, 48,* 1764-1774.

Pickett-Heaps, J. D., West, J. A., Wilson, S. M. and McBride, D. L. (2001). Time-lapse videomicroscopy of cell (spore) movement in red algae. *European Journal of Phycology, 36,* 9-22.

Provasoli, L., McLaughlin, J. J. A. and Droop, M. R. (1957). The development of artificial media for marine algae. *Archiv für Mikrobiologie, 25,* 392-428.

Provasoli, L. (1963). Growing marine seaweeds. In: A. D. De Virville and J. Feldman (Eds.) *Proceedings of the Fourth International Seaweed Symposium* (pp. 9-17). Oxford, Pergamon Press.

Saavedra, L., Balbi, V., Dove, S. K., Hiwatashi, Y., Mikami, K. and Sommarin, M. (2009). *Characterization of phosphatidylinositol phosphate kinase* from the moss *P*hyscomitrella patens*: PpPIPK1 and PpPIPK2. Plant and Cell Physiology, 50,* 595-609.

Saga, N. and Kitade, Y. (2002). *Porphyra*: a model plant in marine sciences. *Fisheries Science, 68 (Suppl II),* 1075-1078.

Saimi, Y. and Kung, C. (2002). Calmodulin as an ion channel subunit. *Annual Review of Physiology, 64,* 289-311.

Sanders, D., Pelloux, J., Brownlee, C. and Harper, J. F. (2002). Calcium at the crossroads of signaling. *Plant Cell, 14,* S401-S417.

Searles, R. B. (1080). The strategy of the red algal life history. *American Naturalist, 115,* 113-120.

Schmid, A. C., Byrne, R. D., Vilar, R. and Woscholski, R. (2004). Bisperoxovanadium compounds are potent PTEN inhibitors. *FEBS Letters, 566,* 35-38.

Smith, R. J., Sam, L. M., Justen, J. M., Bundy, G. L., Bala, G. A. and Bleasdale, J. E. (1990). Receptor-coupled signal transduction in human polymorphonuclear neutrophils: effects of a novel inhibitor of phospholipase C-dependent processes on cell responsiveness. *Journal of Pharmacology and Experimental Therapeutics, 253,* 688-697.

Springer, M. L. (1993) Genetic control of fungal differentiation: the three sporulation pathways of *Neurospora crassa. Bioessays, 15,* 365-374.

Takahashi, M., Saga, N. and Mikami, K. (2010). Photosynthesis-dependent extracellular Ca^{2+} influx triggers an asexual reproductive cycle in the marine red macroalga *Porphyra yezoensis. American Journal of Plant Sciences, 1,* 1-11.

Taylor, A. R., Manison, N. F. H., Fernadez, C., Wood, J. and Brownlee, C. (1996). Spatial organization of calcium signaling involved in cell volume control in the *Fucus* rhizoid. *Plant Cell, 8,* 2015-2031.

ten Hove, C. A. and Heidstra, R. (2008) Who begets whom? Plant cell fate differentiation by asymmetric cell division. *Current Opinion in Plant Biology, 11,* 34-41.

Thiery, J. P., Acloque, H., Huang, R. Y. J. and Nieto, M. A. (2009). Epithelial-mesenchymal transitions in development and disease. *Cell, 25,* 871-890.

Thornber, C. S. (2006). Functional properties of the isomorphic bibasic algal life cycle. *Integrative and Comparative Biology, 46,* 605-614.

Tripodi, G. and de Masi, F. (1983). Unusual structure in the spermatial vesicles of the red alga *Erythrocystis montagnei. Plant Systematics and Evolution, 143,* 197-206.

Tseng, C.K. and Chang, T.J. (1955). Studies on the *life history* of Porphyra *tenera* Kjellm. *Scientia Sinnica, 4,* 375-398.

Wang, J. A., Dai, J. X. and Zhang, Y. T. (2006). Nuclear division of the vegetative cells, conchosporangial cells and conchospores of *Porphyra yezoensis* (Bangiales, Rhodophyta). *Phycological Research, 54,* 201–207.

Wheeler, G. L. and Brownlee, C. (2008). Ca^{2+} signalling in plants and green algae-changing channels. *Trends in Plant Science, 13,* 506-514.

Yoon, G. M., Dowd, P. E., Gilroy, S. and McCubbin, A. G. (2006). Calcium-dependent protein kinase isoforms in *Petunia* have distinct functions in pollen tube growth, including regulating polarity. *Plant Cell, 18,* 867-878.

In: Porphyra yezoensis
Editor: Koji Mikami

ISBN 978-1-61122-304-0

Chapter 3

CHLOROPLAST TRANSCRIPTIONAL MACHINERY OF RED ALGAE: CONSERVATION OF FOUR TYPES OF TRANSCRIPTIONAL REGULATORS IN NON-GREEN CHLOROPLASTS

Kan Tanaka
Graduate School of Horticulture, Chiba University
468 Matsudo, Matsudo, Chiba 271-8510, Japan

ABSTRACT

Structural analysis of chloroplast genomes from non-green plant lineages revealed that bacteria-type RNA polymerase subunits and several types of transcriptional regulators are encoded therein as derived from the cyanobacterial endosymbiont. Typically, four transcription factors, designated Ycf27, Ycf28, Ycf29 and Ycf30, as well as a two-component histidine sensory kinase Ycf26, are found in chloroplast genomes of red algal lineage as *Porphyra*, which appears to have been partially or completely lost in other algal and land plant lineages. In this article, we summarize our present knowledge on evolution, structure and function of these chloroplast-encoded regulatory components and the relevant transcription machinery, together with related nuclear-encoded factors.

INTRODUCTION

Chloroplast, an organelle found in eukaryotic plant cells, originated from the endosymbiosis of a cyanobacteria-like photosynthetic bacterium more than one billion years ago. Chloroplasts still retain their own DNA genomes even now, and those genomes are replicated and the encoded genetic information is transcribed and translated as in ancestral bacteria within the chloroplast compartment independent of the cytosolic eukaryotic system.

Author for correspondence: e-mail kntanaka@faculty.chiba-u.jp

Sequence analyses of chloroplast genomes from divergent plant lineages revealed that the encoded genes have high similarity to cyanobacterial counterparts, which indicated the close relationship between the chloroplast ancestor and cyanobacteria. Since various characteristics of chloroplasts, such as photosynthetic pigments and structure of light-harvesting complexes, are rather divergent among plant lineages, it was not clear whether the various chloroplasts are monophyletic or have independent origins. However, extensive phylogenetic analyses revealed that chlorophyte, rhodophyte and glaucophyte were monophyletic and originated from the unique primary endosymbiosis, and that other plant lineages resulted from secondary or further endosymbiosis by eukaryotic (in most cases red or green) algae into non-photosynthetic eukaryotes.

In every case, it was found that the number of chloroplast-encoded genes is very few as compared with the speculative ancestor's genome. After the symbiosis, many endosymbiont's genes are likely to have been simply lost, or transferred to the nuclear genome, and the resultant nuclear-encoded chloroplast proteins are now synthesized in the cytoplasm and imported into chloroplasts after the translation. Thus, protein complexes involving various chloroplast functions as two photosystems are made by coordination of two independent, the chloroplast and the nuclear genomes. The transcriptional machinery, RNA polymerase, and the other transcriptional regulatory components are also the case for such coordination of two genomes, and are dealt with in this article. Other than the RNA polymerase, bacterial transcription machinery is composed of additional regulatory proteins called transcription factors, which are connected to various signal-transduction and environmental-sensing systems. Chloroplast lineages of non-green plants retain typically four-types of these transcription factors, as in red algae, encoded by their chloroplast genomes. These factors and relevant regulatory systems appear to protect the photosynthetic systems robustly as four guardians in the rapidly changing environment. In this review, I summarize our knowledge about their physiological significance.

CHLOROPLAST TRANSCRIPTION MACHINERY

In general, bacterial genomes are transcribed by a single RNA polymerase composed of several subunits, α, β, β', ω and σ. α, β, β' and ω subunits make a tightly bound core complex, $\alpha_2\beta\beta'\omega$, which retains the RNA synthesis catalytic activity. The σ subunit enables the core to recognize promoter sequence and to initiate transcription, and importantly in most bacterial cells, has several molecular species. Thus, the substitution of σ subunits makes possible the unique core RNA polymerase to change the promoter recognition specificity. In addition to σ subunits that make RNA polymerase heterogeneity, transcription factors, which bind to specific promoter DNA sequence and modulate the promoter activity in response to various environmental cues, are playing critical roles for the bacterial transcriptional regulation.

Among divergent bacterial lineages, β' subunit of the core RNA polymerase is split into two parts, N-terminal γ and C-terminal β' subunits, respectively, only in cyanobacterial lineage. This particular characteristic is also observed in chloroplast genome-encoded RNA polymerase, indicating the close relationship between these two RNA polymerases and genetic systems. The chloroplast σ subunit has multiple forms in most cases as in bacteria, but they are always encoded by the nuclear genome. Subunits for the core enzyme, α, β, β' (γ in cyanobacteira), β'' (β' in cyanobacteria) are basically encoded by the chloroplast genome, but gene for the ω subunit has been identified only in some chloroplast genomes. It is not clear whether ω of other plants have been lost during the evolution or transferred to the nuclear genome. As another exception, α subunit is missing from the chloroplast genome and encoded by the nuclear genome in a moss *Physcomitrella patens* [Kabeya et al., 2007].

UNEXPECTED COMPLEXITY OF RNA POLYMERASE IN VASCULAR PLANT CHLOROPLASTS

Thus far, detailed functional analyses for chloroplast RNA polymerase have been performed mainly for vascular plant species. Specific gene disruption technology for chloroplast gene was developed in *Nicotiana tabacum* and used to knock out the chloroplast *rpoB* gene for the RNA polymerase β subunit [Allison et al., 1996]. The resultant mutant plant showed albino phenotype, showing that the RNA polymerase is essential for the chloroplast biogenesis. While most transcripts for chloroplast photosynthetic genes have disappeared in the mutant, transcripts for housekeeping genes and others were still transcribed and rather increased, indicating the presence of another type of RNA polymerase working in the chloroplast [Hajdukiewicz et al., 1997]. Subsequent analyses revealed, in brief, that single-subunit type (T7 bacteriophage-type) RNA polymerase originated from the mitochondrial RNA polymerase was evolved in vascular and other land plants and targeted into chloroplasts [Hedtke et al., 1997]. Since this RNA polymerase is encoded by the nuclear genome, this type of RNA polymerase was named NEP (Nuclear-encoded plastid RNA polymerase), whereas bacteria-type conventional RNA polymerase designated PEP (Plastid-encoded plastid RNA polymerase). Complex interaction and regulation of these two types of RNA polymerase are likely critical for the complex biogenesis and maintenance of various plastids in the vascular plant system.

Functional analyses of multiple sigma factors for PEP have been reported mostly in vascular plants. Nuclear genome analysis in *Arabidopsis* revealed the presence of six sigma factor genes for PEP, Sig1-6 [Tanaka et al., 1997; Fujiwara et al., 2000]. All the predicted gene products are structurally categorized in the primary sigma factors responsible for the consensus-type promoters, and presently no evidence for minor group-3 type sigma factors for chloroplast transcription. Thus far, identified sigma factors from vascular plants are orthologous to either of these *Arabidopsis* sigma factors, and vascular plants appear to be rather homogeneous as for the sigma factor composition. Thus, *Arabidopsis* could be a model experimental system for the PEP functional analysis. With respect to the function of each sigma factor, molecular genetics analyses have been performed using tagged mutant lines in *Arabidopsis* and rice [Kanamaru and Tanaka, 2004]. Especially related to this review, Sig5 is responsive to various stresses, such as high light, salt and low temperature, and activates blue light responsive promoter (BLRP) of *psbD* encoding the reaction center protein D2 of photosystem II [Nagashima et al., 2004; Tsunoyama et al., 2004].

Other than the transcriptional machineries inherited from the ancestral bacterium, chloroplast RNA polymerase of vascular plants has been identified as tightly bound to DNA, and called TAC (transcriptionally active chromosome). Proteomics analyses revealed protein components of TAC, pTAC proteins. They include PPR proteins and other unknown proteins [Schröter et al., 2010], which are not found in red algal chloroplasts, suggesting the evolution of these proteins specifically among land plant lineages (Figure 1).

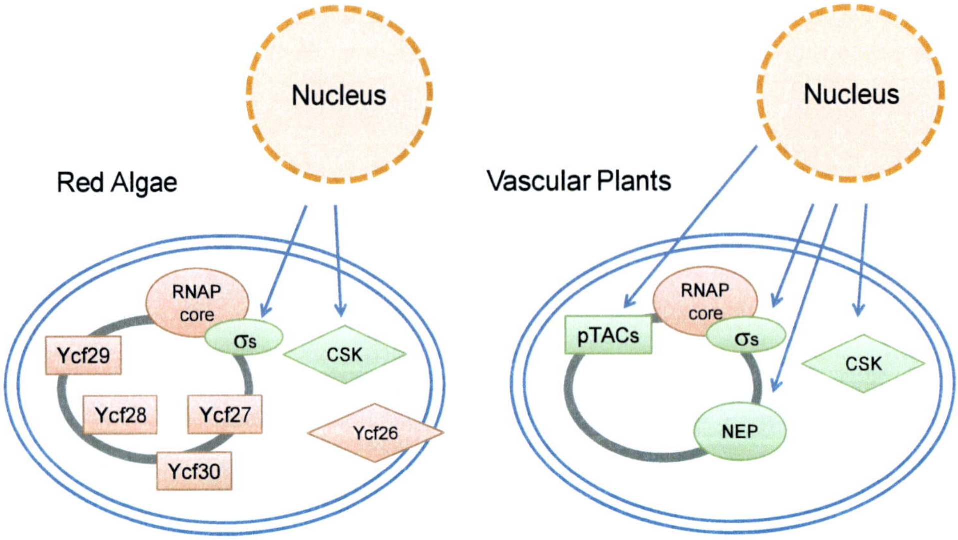

Figure 1. Comparison of typical chloroplast transcription machinery between red algae and vascular plants. In the red algal system, RNA polymerase core subunits are encoded by the chloroplast genome, while sigma subunits and a two-component sensory kinase are nuclear-encoded. In addition, four types of transcription factors and one sensory kinase are chloroplast-encoded. In vascular plants, while bacteria-type RNA polymerase has the similar composition, the RNA polymerase (PEP) is found in TAC making complex with pTAC proteins. Bacteriophage-type RNA polymerase (NEP) is additionally involved in the chloroplast transcription, but chloroplast-encoded transcription factors and the sensory kinase have disappeared, whereas the nuclear-encoded kinase (CSK) is conserved. Red and green colors represent chloroplast-encoded and nuclear-encoded components, respectively.

Chloroplast RNA Polymerase of Red Algae

Information about the chloroplast transcription machinery of red algae was obtained from the chloroplast genome sequences in multicellular red algae as *Porphyra* [Reith and Munholland, 1993]. Subunits for bacteria-type RNA polymerase, α, β, β', β'' and ω were found from the chloroplast genome, encoded by *rpoA*, *rpoB*, *rpoC1*, *rpoC2* and *rpoZ* (or *ycf61*) genes, respectively, indicating the presence of PEP-type enzyme. No σ factor gene was found in any chloroplast genome. Although no functional evidence was provided, at least one cDNA clone, AV432534, shows strong similarity with bacterial principal-type sigma factors, which was found in *Porphyra yezoensis* EST index in the KDRI (Kazusa DNA research institute) database (http://est.kazusa.or.jp/en/plant/porphyra/EST/index.html). Thus, the basic scheme of chloroplast RNA polymerase is likely the same as green plant lineages as for the nuclear encoding of the chloroplast sigma factor. Chloroplast RNA polymerase appears to be similar also in unicellular red algae as *Cyanidium caldarium* [Glöckner et al., 2000] and *Cyanidioschyzon merolae* [Ohta et al., 2003], except that RpoC is not divided into two parts, RpoC1 and RpoC2, in *C. merolae*. This characteristic is only found for *C. merolae* even in related algae, and therefore, mutation(s) resulting in the fusion of *rpoC1* and *rpoC2* genes may have occurred in *C. merolae* after the divergence from other species.

As the nuclear-encoded sigma factor genes, a nuclear gene *sigA* was identified in *Cyanidium caldarium* RK-1, as the firstly found chloroplast sigma factor gene in the plantae [Tanaka et al., 1996]. Subsequently, two nuclear genes encoding chloroplast sigma factors, *sigB* and *sigC*, were identified in the same strain [Oikawa et al., 1998]. Northern blot analysis indicated that expression levels of mRNA of *sigB* and/or *sigC* were increased by the light illumination, whereas the *sigA* mRNA was constitutively found irrespective of the light condition. However, *sigB* and *sigC* could not be discriminated in these experiments because of their very high similarity to each other. In addition, the complete nuclear genome information is available in *C. merolae*, and the bioinformatics analysis revealed the presence of four sigma factor genes, Sig1-4 [Matsuzaki et al., 2004]. These data show that chloroplast sigma factors have multiple forms in red algae as in green plants. The *C. merolae* nuclear genome analysis also made clear that there is only one gene for the mitochondrial RNA polymerase, which is unlikely targeted into the chloroplast. Although some nuclear gene products are likely dual targeted to both chloroplast and mitochondrion, thus far, there is no supporting evidence for the bacteriophage-type chloroplast RNA polymerase (NEP) in red algal chloroplasts.

Four-Types of Transcription Regulators Encoded by Chloroplast Genomes

Other than genes for RNA polymerase, chloroplast genomes of red algae encode four types of bacteria-type transcription factors, Ycf27-30, which are considered to function for the environmental acclimation by modulating the chloroplast gene expression. The comparison of chloroplast transcription machineries between red algae and vascular plants is summarized in Figure 1. Other than red algae, various algal lineages originated from red algal secondary endosymbioses also have descended some of these transcription factors [Douglas and Penny, 1999; Jacobs et al., 1999; Glöckner et al., 2000; Duplessis et al., 2007], and the distribution among lineages is summarized in Table 1. In this review, I describe our current understanding of their function and the physiological relevance.

Chloroplast Two-Component Systems

Two-component system is the central signal transduction mechanism in bacteria [Galperin, 2006]. Bacteria cells are able to sense changes of environmental parameters, and activate or repress various sets of genes for acclimation. Using versatile bacterial systems as *Escherichia coil* and *Bacillus subtilis*, a number of detailed analyses have been performed for each signaling pathway, and it was found that common regulatory components composed of a kinase and a transcription factor are involved in a number of cases. A specific type of protein kinases is involved here, called "histidine kinase (Hik)" in which a conserved histidine residue (H) is first autophosphorylated and the activated phosphate group is used for the phosphotransfer reaction to the cognate receiver protein called "response regulator (Rre)."

Table 1. Distribution of chloroplast transcriptional regulators among plant lineages

	CSK	Ycf26	Ycf27	Ycf28	Ycf29	Ycf30	Others
Rhodophyte							
Porphyra purpurea	NE	+	+	+	+	+	
Porphyra yezoensis	NE	+	+	+	+	+	
Gracilaria tenuistipitata	NE	+	+	+	+	+	
Cyanidioschyzon merolae	+	-	+	+	+	+	
Cyanidium caldarium	NE	+	+	+	+	+	
Cryptophyte							
Gillardia theta	NE	-	+	-	+	+	
Haptophyte							
Emiliania huxleyi	NE	+	+	-	-	+	
Raphidophyte							
Heterosigma akashiwo	+	+	+	-	-	+	
Diatom							
Thalassiosira pseudonana	+	-	-	-	-	+	
Glaucophyte							
Cyanophora paradoxa	NE	-	+	-	+	+	
Chlorophyte							
Arabidopsis thaliana	+	-	-	-	-	-	
Chlorokybus atmosphyticus	NE	-	+	-	-	+	
Rhizaria							
Paulinella chromatophora	(+)	+	++	-	+	+	+++++

This table was made based on available information in references [Douglas and Penny, 1999; Duplessis et al., 2007; Oudot-Le Secq et al., 2007; Puthiyaveetil and Allen, 2009] and genome databases (http://www.ncbi.nlm.nih.gov/sites/entrez?db=genome; http://merolae.s.u-tokyo.ac.jp/). The number of "+" indicates the number of paralogs. Note that CSK is encoded by the chromatophore genome in *Paulinella*. NE: not examined.

Response regulator usually function as transcription factors whose conserved aspartic acid residue (D) is phosphorylated to change the protein conformation, resulting in the modulation of transcription from the corresponding target promoters of a set of genes for specific roles. Examples of responses by two-component system are for osmolarity, phosphate starvation, nitrogen starvation in *E. coli*, and sporulation initiation in *B. subtilis*.

Recent microbial whole genome analyses revealed dozens of two-component systems are working in parallel, though there are some cross-talks among the phosphorylation reactions, and constituting the predominant signal transduction system for environmental acclimation. This is also the case in cyanobacteria, and 44/42 and 13/24 Hik/Rre are encoded by the *Synechocystis* sp. PCC 6803 [Kaneko et al., 1996; Mizuno et al., 1996] and *Synechococcus elongatus* PCC 7942 genomes, respectively. These two cyanobacterial strains have been the model strains to understand the cyanobacterial cell regulation, and therefore, I mainly refer to these two cyanobacteria in this review in comparison with chloroplast systems. Multiple Hiks and Rres are designated as numbers, such as Hik1, Hik2, and Rre1, Rre2, in *Synechocystis* PCC 6803, and usually in more conventional names in *S. elongatus* PCC 7942.

While chloroplast genomes of the green lineage do not encode for two-component regulators, with an exception as I discuss below, two of four transcription factors in red lineages, Ycf27 and Ycf29, are response regulators of bacterial two-component system. As the relevant Hik encoding gene, one is found in the chloroplast genome, and another was recently found in the nuclear genome.

Chloroplast Response Regulators, Ycf27 and Ycf29

Ycf27: Ycf27 is an Rre that is categorized in the OmpR family [Galperin, 2006]. This type of Rres is composed of the N-terminal receiver domain containing the aspartic acid residue for the phosphorylation and the C-teminal winged-helix type DNA binding domain. Functional analysis of chloroplast Ycf27 is thus far scarce, partly because reverse genetics of the red chloroplast genome has been presently difficult while described in a paper [Lapidot et al., 2002]. With respect to the gene arrangement in the chloroplast genome, *ycf27* gene is conservatively positioned at the upstream of *psbDC* operon in the opposite direction. Giving that the *psbD* gene encodes the D2 protein of photosystem II reaction center, and that the *psbD* transcription is regulated in response to light conditions [Minoda et al., 2005; our unpublished observation], it is conceivable that Ycf27 is involved in the *psbD* promoter activation.

While little information is available for chloroplast Ycf27, functional analyses of cyanobacterial homologs have been reported. In *Synechocystis* sp. PCC 6803, two related homologs of *ycf27*, slr0115 and slr0947, were found, and named *rpaA* and *rpaB*, respectively [Ashby and Mullineaux, 1999]. Subsequent analysis indicated that *rpaB* is orthologous to *ycf27*, while *rpaA* is not [Ashby et al., 2002]. Knock out of *rpaA* resulted in increased efficiency of energy transfer from phycobilisome to photosystem II. The *rpaB* gene was likely essential for the viability, but the reduction of the gene copy number had the opposite phenotypic effect. These results indicate the involvement of these Rres for modulation of photosynthetic energy transfer.

As a more physiological role, the *rpaA* mutant was found to lose circadian-rhythm dependent periodic gene expression in *S. elongatus* PCC 7942 [Takai et al., 2006]. Previously, a Hik, SasA, was shown to interact with the central circadian oscillator protein KaiC, and required to sustain robust oscillation [Iwasaki et al., 2000]. Thus, RpaA is composing a two-component system with SasA in the output pathway of the circadian transcriptional oscillation. RpaB is also conserved in *S. elongatus* PCC 7942 and other cyanobacteria, but its mechanism for the transcriptional regulation have been uncovered only recently. Examining the induction mechanism of high light-responsive gene *hliA* in *S. elongatus* PCC 7942, Kappel et al. identified a cis-acting negative element in the promoter region and named this sequence high light regulatory 1 (HLR1) [Kappel et al., 2006]. They further found the histidine kinase NblS is involved in the negative regulation [Kappel et al.,

2006]. The same group also identified HLR1 sequence in the high light responsive *hliB* promoter in *Synechocystis* PCC 6803, and RpaB was shown to be responsible for the binding [Kappel and van Waasbergen, 2007]. More detailed mechanism for the high-light induction has been proposed recently [Seki et al., 2007; Hanaoka and Tanaka, 2008; Seino et al., 2009]. Under the normal light condition, RpaB protein is kept under the phosphorylated state and bound to HLR1. NblS (Hik33 in *Synechocystis* PCC 6803) is likely involved in the release of RpaB from HLR1 in case of high light stress, which results in the de-repression of high-light responsive genes as *hliA*, *hliB* and *rpoD3* (high-light inducible group-2 sigma factor; orthologous to *sigD* in *Synechocystis* PCC 6803), or inhibits activation of photosystem I genes as *psaAB* [Seino et al., 2009]. Release of RpaB from HLR1 was shown to occur within one minute as shown by the chromatin immunoprecipitation analysis [Hanaoka and Tanaka, 2008]. Thus, the reduction of *rpaB* gene copy number, in the experiment described above [Ashby and Mullineaux, 1999], may have mimicked the high-light response to decrease the energy transfer efficiency from the phycobilisome to photosystem II. These results suggest that Ycf27 of the chloroplast system is also functioning similarly.

Ycf29: Ycf29 is an Rre of NarL-type [Galperin, 2006], composed of the N-terminal receiver domain and the C-terminal typical helix-turn-helix DNA binding domain. As compared with Ycf27, far little information is available about the chloroplast Ycf29 function. Ycf29 is orthologous to Rre1 in *Synechocystis* PCC 6803, and the homolog is highly conserved among cyanobacteria. To understand the signal transduction network composed of two component regulators, knock out mutants for each *hik* and *rre* genes were systematically constructed, and the changes of transcriptomes were analyzed by microarray analyses in *Synechocystis* PCC 6803. In these analyses, transcripts accumulation of several genes, including heat shock protein genes such as *hspA*, *htpG*, *dnaK2* and *groEL,* were decreased in case of salt or hyperosmotic stress by the *rre1* mutation [Shoumskaya et al., 2005, Figure 2]. Giving that *dnaK* and *groEL* genes are also found in the chloroplast genomes of red algae, these genes would be possible candidates for the Ycf29 regulation. Another study showed that an alcohol dehydrogenase is induced by stress conditions under the control of Hik34-Rre1 two-component system [Vidal et al., 2009]. However, the direct relevance is unknown because the corresponding Hik and the target gene is absent from the chloroplast system.

Histidine kinases for chloroplast two-component system and the partnership with response regulators: Since one *hik* gene, *ycf26*, is found in the chloroplast genome of red algae, the encoded Hik is the probable candidate for the kinase partner of two-component Rres, Ycf27 and Ycf29. The predicted protein structure indicates that Ycf26 is a membrane-spanning protein whose kinase domain is located in the cytosolic side. Ycf26 is also highly conserved among cyanobacteria, and especially two orthologs, Hik33 (DspA) in *Synechocystis* PCC 6803 and NblS in *S. elongatus* PCC 7942, have been investigated in detail. Hik33 was first identified as a sensory kinase that is responsible for low-temperature induction of fatty acid desaturate gene *desD* [Suzuki et al., 2001; Suzuki et al., 2000]. Subsequent analyses revealed that Hik33 is not only responsive to low-temperature stress but also to various stress conditions such as hyperosmolarity, salt, and H_2O_2 stresses [Mikami et al., 2002; Marin et al., 2003; Paithoonrangsarid et al., 2004; Shoumskaya et al., 2005; Kanesaki et al., 2007] to activate gene expressions. On the other hand, other reports showed that Hik33 represses high-light responsive genes under low light condition [Tu et al., 2004; Hsiao et al., 2004]. Consistent explanation for the whole view of the Hik33 role would require further studies. The *nblS* gene of *S. elongatus* PCC 7942 was first found as a gene involved in photosynthesis-related gene expression during high-light and nutrient stress [Salinas et al., 2007; van Waasbergen et al., 2002]. The name of *nblS* was designated from the non-bleaching phenotype in case of high-light stress or nutrient deprivation, which indicates the defects to induce *nblA* for the phycobilisome degradation.

Recently, a nuclear-encoded Hik, which is highly conserved among various plant lineages, was found to localize in the chloroplast compartments, and named chloroplast sensory kinase, CSK [Puthiyaveetil and Allen, 2009; Puthiyaveetil et al., 2009]. Since there is no typical chloroplast targeting signal in the N-teminus, the localization mechanism is presently unknown. This Hik appears soluble because of the absence of membrane spanning domain, and contains GAF-domain at the center and the C-terminal kinase domain. GAF-domain is present in phytochromes and cGMP-specific phosphodiesterases, and suggested to involve light signal transduction [Aravind and Ponting, 1997]. CSK is likely highly conserved in the green plant lineage, and also in cyanobacteria as Hik2 orthologs. Nuclear genome analysis of the unicellular red alga *C. merolae* revealed the presence of unique Hik gene (CMP021C) [Matsuzaki et al., 2004], and this gene is the CSK homolog. Thus, CSK is also conserved in red algae, and the chloroplast two-component system would be composed of two Hiks, CSK and Ycf26, and two Rres, Ycf27 and Ycf29. Very peculiarly among red algae, *C. merolae* lacks the chloroplast-encoded *ycf26* gene. Thus, the nuclear-encoded CSK is the sole chloroplast Hik, and likely to take parts in phosphorylation of both response regulators, which implies the cross-talk of two-component regulators in other red algal chloroplasts. In *Synechocystis* PCC 6803, yeast two-hybrid analysis identified interaction between Hik2 and Rre1 [Paithoonrangsarid et al., 2004], and this interaction is very consistent with the probable CSK and Ycf29 interaction in red algal chloroplasts. Two-component regulators orthologous to CSK (Hik2), Ycf26 (Hik33), Ycf27 (RpaB or Re26) and Ycf29 (Rre1) are highly conserved among divergent cyanobacteria, whereas most of other regulators are specific for subgroups [Ashby and Houmard, 2006]. This suggests that these regulators are essential for stress responses for oxygen-evolving photosynthetic organisms. It is of interest that CSK is highly conserved even if other two-component regulators have been lost during the green plant evolution [Puthiyaveetil and Allen, 2009]. This may indicate that some other protein(s) has/have evolved to be phosphorylated by CSK in some contexts. Response regulator-like protein found in a vascular plant chloroplast might be a candidate for this [Weber et al., 2006].

Possible Nitrogen Regulator, Ycf28

ycf28 is predicted to encode a transcription factor of CRP (cAMP receptor protein of *E. coli*) family, and considered to be orthologous to *ntcA* of cyanobacteria. Each cyanobacterial strain has several types of CRP family transcription factors, some of which were actually shown to be the cAMP receptor proteins [Yoshimura et al., 2000]. Phylogenetic analyses of Ycf28 proteins showed that most probable orthologs of cyanobateria are NtcA, but further experimental evidence should be required to make clear the orthologous relationship.

NtcA was first identified as the central nitrogen regulator in a cyanobacterium, *S. elongatus* PCC 7942, and positively regulates nitrogen assimilatory genes, such as glutamine synthetase (GS) and transporters for inorganic nitrogen, in response to environmental nitrogen depletion [Vega-Palas et al., 1992]. NtcA binds to the conserved DNA sequence “GTA-N_8-

TAC" located in the upstream region of the target promoters, which is considered to directly activate RNA polymerase. NtcA is globally conserved among cyanobacteria and shown to be responsible for most physiological responses to nitrogen-starved conditions. In cyanobacteria, tricarbonate (TCA) cycle is not complete because of the lack of 2-oxoglutarate dehydrogenase, and 2-oxoglutarate (2-OG) is produced by the right half of the TCA cycle and consumed predominantly by the nitrogen assimilation pathway composed of GS and glutamate synthase (glutamate 2-oxoglutarate aminotransferase; GOGAT), GS-GOGAT cycle. Thus, 2-OG rapidly accumulates under nitrogen depletion because of the decline of the GS-GOGAT reaction. Since the change of the intracellular 2-OG appeared to correlate with the transcriptional activation of nitrogen assimilatory genes, involvement of 2-OG for the NtcA activity was expected, and subsequently it was found that 2-OG enhanced the NtcA-DNA interaction and activated NtcA-dependent promoter transcription [Tanigawa et al., 2002; Vazquez-Bermudez et al., 2002; Zhao et al., 2010].

The physiological role and the target genes of Ycf28 are almost unknown. In case of cyanobacteria, a number of genes have been identified under the control of NtcA. However, most of such target genes are not found in the non-green chloroplast genome. Subunits of GOGAT are chloroplast-encoded as components for nitrogen assimilation, but appear not under the control of NtcA in cyanobacteria and not activated by nitrogen depletion in the chloroplast of a most intensively studied unicellular red alga, *C. merolae* (our unpublished result). Only two genes, *glnB* and *nblA*, which have been well known as NtcA-regulated genes in cyanobacteria, are found in typical red algal chloroplast genomes. The *glnB* gene encodes PII protein that is a widely conserved regulator for carbon-nitrogen assimilation balance. PII is a small protein of about 110 amino acids, and usually found as a homotrimer. Direct binding to 2-OG and/or ATP changes the protein conformation, and thus the interaction with various target proteins, which result in the modulation of transcription factor and metabolic enzyme activities. Among them, well-characterized targets are PipX to regulate the NtcA activity in *S. elongatus* PCC 7942 [Espinosa et al., 2006] and N-acetyl-l-glutamate kinase (NAGK) for arginine biosynthesis in *S. elongatus* PCC 7942 and green plant chloroplasts [Burillo et al., 2004; Maheswaran et al., 2004; Sugiyama et al., 2004; Ferrario-Mery et al., 2006]. In *Porphyra* and other red algae, since *argB* and *glnB* genes for NAGK and PII, respectively, are both found in the chloroplast genome, it may constitute a regulatory network together with Ycf28. An exception is *C. merolae*, in which *glnB* gene was not found in the chloroplast genome. In addition, PII protein is encoded by the nuclear genome in vascular plants, but it is thus far not identified even from the nuclear genome in *C. merolae*.

Another candidate for the Ycf28 target gene in chloroplasts is *nblA* (or *ycf18*). The *nblA* gene was identified as a factor required for the phycobilisome (PBS) degradation in case of nitrogen depletion or high light stress, and named after non-bleaching phenotype of the mutant. NblA, a small protein of about 60 amino acids, interacts with PBS as a homotrimeric form, which is considered to be essential for the PBS degradation. Transcriptional activation of *nblA* by nitrogen depletion was shown strong dependence on NtcA in cyanobacteria. Thus, because PBS degradation by nitrogen depletion is also observed in red algal chloroplasts and the *nblA* homolcg gene (*ycf18*) is found in the chloroplast genomes, transcriptional regulation of *ycf18* was examined in P. yezoensis. With regard to the results, the authors found that *ycf18* was induced by ammonium addition rather than nitrogen starvation, suggesting that *ycf18* has a role other than the PBS degradation [Kawakami et al., 2009]. Our experiments using *C. merolae* are also negative for the *ycf18* induction by nitrogen starvation (unpublished observation), and thus the role of *ycf18* should require further study to be determined.

Overall scheme of nitrogen assimilation in red algae has been best understood in a unicellular *C. merolae* [Imamura et al., 2010]. Unique GS is encoded by the nuclear genome, and localized in the cytoplasm [Terashita et al., 2006]. GOGAT appears to be only in the chloroplast because the gene is encoded by the chloroplast genome, and thus, once assimilated, nitrogen should be imported into the chloroplast compartment as glutamine for the GOGAT reaction to produce glutamate. Another substrate for the GOGAT reaction, 2-OG, is likely produced by the mitochondrion as in other plant systems, which once exported to the cytoplasm and subsequently imported into the chloroplast. In case of nitrogen starvation, nuclear-encoded nitrogen assimilatory genes are activated by an R2R3-type Myb transcription factor MYB1 [Imamura et al., 2009], but the upstream signal transduction mechanisms are not known presently. Given that Ycf28 is orthologous to cyanobacterial NtcA, Ycf28 is likely a sensor of 2-OG in the chloroplast compartment, and thus the metabolic conditions of the cytoplasm and the chloroplast might be sensed independently. Sensing and coordination of nitrogen starvation signal is not known in plant cells in general, and future study using *C. merolae* would result in our deeper understanding of nitrogen regulation in plant cells.

RubisCO Regulator, Ycf30

Ycf30 is a LysR-type transcription regulator (LTTR) [Maddocks and Oyston, 2008; Schell, 1993], which is homologous to the widely conserved regulator for RubisCO genes among cyanobacteria, photosynthetic bacteria and chloroplasts of red lineages. This type of transcription factor harbors helix-turn-helix DNA binding domain at the N-terminus, while the C-terminal domain is able to recognize and binds specific low molecular weight co-inducer, binding of which results in the protein conformation changes for the transcriptional modulation. For example, a landmark study by Stragier et al. showed that diaminopimelate decarboxylase gene is regulated by binding of co-inducer, diaminopimelate, to the typical LTTR, LysR in *E. coli* [Stragier et al., 1983].

In green plant lineages, two subunits of RubisCO are separately encoded by the nuclear (*rbcS*) and the chloroplast (*rbcL*) genomes, and no bacteria-type transcriptional regulator has been identified with an exception (see below), which appears to have been lost during the evolution. These *rbc* genes are phylogenetically closely related to those of the cyanobacterial endosymbiont, while the *rbcS* gene was likely transferred to the nuclear genome subsequently. In contrast, both *rbcS* and *rbcL* genes are encoded by the chloroplast genome as an operon, together with its cognate transcriptional activator gene, *ycf30* (or *rbcR*), in red lineages. Phylogenetic analysis of Ycf30 revealed that this protein is monophyletic with the counterpart of cyanobacteria (for example, Sll0998 of *Synechocystis* PCC 6803), which is consistent with the cyanobacterial origin of the chloroplast [Maier et al., 2000]. However, the *rbcLS* genes are rather related to *cbbLS* of α-proteobacteria and also constitute an operon with *cbbX* (or *cfxQ*) that is conserved among photosynthetic bacteria [Gibson and Tabita, 1997]. Thus, this operon structure should have originated from the horizontal gene transfer event. The details of such evolutional process are not well understood presently. With respect to the *cbbX* gene, the *C. merolae* genome analysis revealed that another *cbbX* gene is encoded as a putative chloroplast protein by the nuclear genome. In a cryptophyte alga *Guillardia theta*, there is a remnant of the nucleus of the sencondary endosymbiotic red alga, nucleomorph, in the chloroplast endoplasmic reticulum. Each of the chloroplast and the nucleomorph genomes of *Guillardia theta* encodes *cbbX* gene [Douglas and Penny, 1999]; therefore, two types of CbbX proteins encoded by the separate genomes may be a common characteristic of red algae. The CbbX protein is usually considered to be a protein required for the RubisCO complex assembly [Maier et al., 2000]. However, recent reports postulate that CbbX has an ability to bind specific DNA sequence and could work as a transcriptional regulator [Fujita et al., 2008a; Fujita et al., 2008b]. Future analysis would make clear these possibilities. Transcript analysis of the chloroplast genes in *C. merolae* revealed that the *rbcLS* operon is induced under the CO_2-limited condition (our unpublished observation). The underlying regulatory system, including the involvement of Ycf30 and/or CbbX, should be further examined as the next step.

Regulatory aspects of these RubisCO regulators have been investigated mostly in photosynthetic bacteria. Similar to LTTR proteins that bind to a consensus sequence, TTA-$N_{7/8}$-TAA, and RbcR (CbbR) proteins are also able to bind these sequences found around the upstream of *rbc* operon promoters, and activate the promoter. In a photosynthetic non-sulfur purple bacteria, *Rhodobacter sphaeroides*, ribulose-1,5-bisphosphate was found to increase the electrophoretic motility of the CbbR/*cbb* promoter DNA complex [Dangel et al., 2005]. Also in *Xanthobacter flavus*, NADPH (but not NADH) was shown to enhance the CbbR-DNA binding [van Keulen et al., 1998]. In a cyanobacterium *S. elongatus* PCC 7942, bicarbonate transporter genes are induced under CO_2-limited condition dependent on RbcR-related transcription factor CmpR [Nishimura et al., 2008]. In this case, ribulose-1,5-bisphosphate (>1 mM) or 2-phosphoglycolate (>10 μM) could enhance the protein-DNA binding, for which the authors suggested that 2-phophoglycolate could be the co-inducer for the transcriptional activation. And finally, it was recently shown that Ycf30-dependent *rbcLS* promoter could be activated by ribulose-1,5-bisphosphate, 3-phosphoglycerate and NADPH using isolated chloroplast system in *C. merolae* [Minoda et al., 2010]. All of these candidates for co-inducer are related to the Calvin cycle product or the substrate, and, thus, would be suitable as the RubisCO operon regulators. As is in prokaryotic system as photosynthetic bacteria and cyanobacteria, monitoring the metabolic flux of the Calvin cycle is probably enough to acclimate to the environmental changes of CO_2 availability. However, in eukaryotic plant systems, Calvin cycle is working in the chloroplast, but the metabolic flux is supported by dynamic traffic of compounds, which are transported through organelles. Genes for photosynthetic reactions are encoded separately by two independent chloroplast and nuclear genomes, both of which are able to respond to environmental CO_2 conditions (our unpublished observation in *C. merolae*). Thus, the inter-organelle communication is likely playing a major role for the overall carbon regulatory system as in the nitrogen regulation described above.

CHLOROPLAST-ENCODED TRANSCRIPTION FACTORS IN NON-RED LINEAGES

Chloroplasts of red algae have a prototype of chloroplast transcriptional regulation of red lineages, and the inherited regulation is also observed in chloroplasts originated from the secondary endosymbiosis of red algae. On the other hand, bacteria-type transcription factors are also found in some chloroplast genomes from non-red lineages. In this section, I pass through these systems briefly and discuss the evolutional implications.

Chlorokybus: Chlorophytes, rhodophytes and glaucophytes are considered to have originated from the unique endosymbiotic event called the primary endosymbiosis. After the diversification of these three major clades, chloroplasts of chlorophytes have been assumed to have completely lost the bacterial transcription factors during the early evolution, because of no trait of such gene repertoires. However, recent sequence analysis of a streptophyte, *Chlorokybus atmophyticus*, revealed that the chloroplast genome encodes for two of the four transcription factors found in red algae, Ycf27 and Ycf30 (RbcR) [Lemieux et al., 2007]. It should be noted that this *ycf27* gene is located adjacent to the 5' upstream of the *psbDC* operon, which is the same arrangement as in red algae, suggesting that this gene is a *bona fide* functional ortholog of the *ycf27* of red algae. If the presence of the gene is not resulted from a horizontal gene transfer, streptophytes are also considered to have retained bacterial transcription factors until the time of current divergence. Providing that the chloroplast regulatory system of *C. atmophyticus* represents the prototype of those of streptophytes, the presence of *ycf27* is quite interesting from the viewpoint of evolution of the chloroplast stress response. As inferred from the cyanobacterial context, *ycf27* is involved in the stress responses especially for the *psbDC* operon in chloroplasts. On the other hand, the stress responsive transcriptional activation of *psbDC* is dependent on the nuclear-encoded chloroplast RNA polymerase sigma factor Sig5 in moss and vascular plants, which does not have Ycf27 [Nagashima et al., 2004; Tsunoyama et al., 2004; Ichikawa et al., 2008]. Thus, the substitution of the regulatory element might have made a large evolutional step, probably associated with the emergence of the land plants. While the *rbcR* (*ycf30*) gene is conserved as in red algae, the chloroplast *rbc* operon of *C. atmophyticus* does not include the *rbcS* gene. The chloroplast genomes of red algae contain the *rbcLS* operon of photosynthetic bacteria origin; therefore, *C. atmophyticus* may retain the original version of the *rbc* gene sets, in contrast to the red algal *rbc* genes transferred from photosynthetic bacteria.

Cyanophora paradoxa: C. paradoxa belongs to glaucophyte that is one of the three major branches originated from the primary endosymbiosis. The complete genome sequence of cyanelle, the *C. paradoxa* chloroplast, was determined, and is available at NCBI database (http://www.ncbi.nlm.nih.gov/sites/entrez?Db=genomeandCmd=ShowDetailViewandTermToSearch=10602). Among the four transcription factors of red algae, *ycf27, ycf29* and *ycf30* genes were found, but the functional analysis has not been performed yet.

Independent chloroplast evolution, chromatophore of Paulinella: Paulinella is a thecate, filose amoeba containing two blue-green photosynthetic inclusions termed chromatophores. The chromatophore contains its own genome, and the sequence analysis made clear that the chromatophore was resulted from an independent primary photosynthetic organelle acquisition from a prey cell related to extant marine cyanobacteria as *Prochlorococcus*/*Synechococcus* [Marin et al., 2005; Yoon et al., 2006; Bhattacharya et al.,

2007; Bodyl et al., 2007; Nowack et al., 2008; Nakayama and Ishida, 2009; Yoon et al., 2009]. Compared with the primary endosymbiosis dated more than a billion years ago, acquisition of the chromatophore occurred relatively recently (minimum age of about 60 million years ago), which is consistent with the fact that the genome size is over 1 megabase pairs containing larger number of genes than other chloroplasts. The chromatophore genome contains genes for all of RNA polymerase subunits, including two sigma factors as unusual as the chloroplast genome. Based on the similarity search analysis, one of the two sigma factors is the group 1 principal sigma, and another is related to the stress responsive sigma factors as classified in M4 cluster in previous literature [Osanai et al., 2008]. Three of the four conserved chloroplast transcription factors are also found in the chromatophore, i.e., Ycf27, Ycf29 and Ycf30, where another paralogous gene for Ycf27 and several other types of transcription factors are also found herein. In addition, it is of particular interest that two Hiks, homologs of Hik2 and Hik33 from *Synechocystis* PCC 6803, are found therein. As described above, these sensory kinases are also found as two-component regulators in red algal chloroplasts. The conservation of almost the same regulatory elements in independently evolved chloroplasts, as well as in diverged cyanobacteria, indicates that these proteins are the "core" part of transcriptional network in cyanobacteria-type photosynthetic systems.

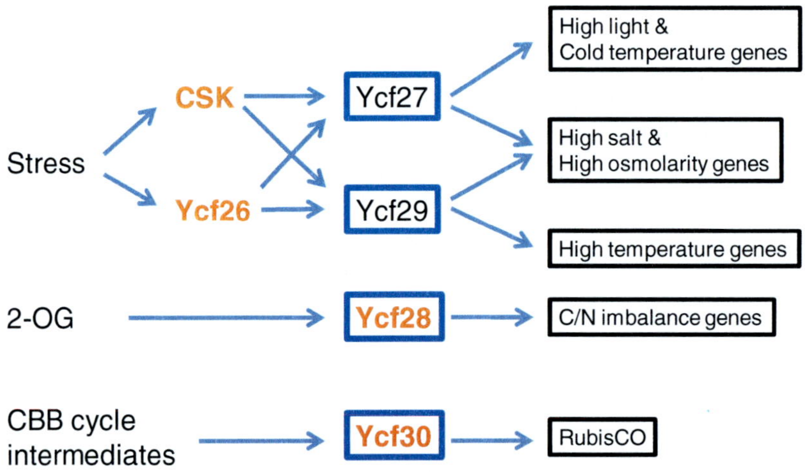

Fgure 2. Function of four transcription factors in red algal chloroplasts. Relevant signals, signal receptor proteins, and the target genes are schematically represented. Blue boxes indicate the four transcription factors directly bind to DNA elements. Red letters show the component receiving each signal. CSK and Ycf26 are sensory kinases for stresses, and phosphorylate either or both response regulators. Note that this scheme was compositely made based on results from various organisms, and includes much speculation to be experimentally examined.

Conclusion

While chloroplast genome of red algae encodes for four types of transcription factors descended from their common ancestor, our understanding for the physiological function and the regulatory mechanism remains still in infancy. On the other hand, homologous regulatory system is found in cyanobacteria, the sister organisms of chloroplasts. Detailed genetic and biochemical analysis can be applied for cyanobacteria, and comparative study using cyanobacteria and red algae have given us many insights into their function in red algae. Also, comparative studies among divergent chloroplast lineages have been useful to understand the framework of environmental acclimation in photosynthetic organisms. Currently, our knowledge about the functions of the four transcriptional regulators is represented in Figure 2, although the large part of the scheme includes much speculation. Keeping these evolutional points of view in mind, and of course together with the developments of versatile experimental systems of red algae as in *C. merolae*, research of the regulatory systems in red algal chloroplasts could play major roles for the understanding of the whole view of photosynthetic regulatory systems.

REFERENCES

Allison, L. A., Simon, L. D., and Maliga, P. (1996) Deletion of *rpoB* reveals a second distinct transcription system in plastids of higher plants. *EMBO J., 15*, 2802-2809.

Aravind, L, and Ponting, C. P. (1997) The GAF domain: an evolutionary link between diverse phototransducing proteins. *Trends Biochem. Sci., 22*, 458-459.

Ashby, M. K., and Houmard, J. (2006) Cyanobacterial two-component proteins: structure, diversity, distribution, and evolution. *Microbiol. Mol. Biol. Rev., 70*, 472-509.

Ashby, M. K., Houmard, J., and Mullineaux, C. W. (2002) The *ycf27* genes from cyanobacteria and eukaryotic algae: distribution and implications for chloroplast evolution. *FEMS Microbiol. Lett., 214*, 25-30.

Ashby, M. K., and Mullineaux, C. W. (1999) Cyanobacterial *ycf27* gene products regulate energy transfer from phycobilsomes to photosystems I and II. *FEMS Microbiol. Lett., 181*, 253-260.

Bhattacharya, D., Archibald, J. M., Weber, A. P., and Reyes-Prieto, A. (2007) How do endosymbionts become organelles? Understanding early events in plastid evolution. *Bioessays, 29*, 1239-1246.

Bodyl, A., Mackiewicz, P., and Stiller, J. W. (2007) The intracellular cyanobacteria of *Paulinella chromatophora*: endosymbionts or organelles? *Trends Microbiol., 15*, 295-296.

Burillo, S., Luque, I., Fuentes, I., and Contreras, A. (2004) Interactions between the nitrogen signal transduction protein PII and N-acetyl glutamate kinase in organisms that perform oxygenic photosynthesis. *J. Bacteriol., 186*, 3346-3354.

Dangel, A. W., Gibson, J. L., Janssen, A. P., and Tabita, F. R. (2005) Residues that influence in vivo and in vitro CbbR function in *Rhodobacter sphaeroides* and identification of a specific region critical for co-inducer recognition. *Mol. Microbiol., 57*, 1397-1414.

Douglas, S. E., and Penny, S. L. (1999) The plastid genome of the cryptophyte alga, *Guillardia theta*: Complete sequence and conserved synteny groups confirm its common ancestry with red algae. *J. Mol. Evol., 48*, 236-244.

Dangel, A. W., Gibson, J. L., Janssen, A. P., and Tabita, F. R. (2005) Residues that influence *in vivo* and *in vitro* CbbR function in *Rhodobacter sphaeroides* and identification of a specific region critical for co-inducer recognition. *Mol. Microbiol., 57*, 1397-1414.

Duplessis, M. R., Karol, K. G., Adman, E. T., Choi, L. Y. S., Jacobs, M. A., and Cattolico, R. A. (2007) Chloroplast His-to-Asp signal transduction: a potential mechanism for plastid gene regulation in *Heterosigma akashiwo* (Raphidophyceae). *BMC Evol. Biol., 7*, 70.

Espinosa, J., Forchhammer, K., Burillo, S., and Contreras, A. (2006) Interaction network in cyanobacterial nitrogen regulation: PipX, a protein that interacts in a 2-oxoglutarate dependent manner with PII and NtcA. *Mol. Microbiol., 6*, 457-469.

Ferrario-Mery, S., Besin, E., Pichon, O., Meyer, C., and Hodges, M. (2006) The regulatory PII protein controls arginine biosynthesis in *Arabidopsis. FEBS Lett., 580*, 2015-1020.

Fujita, K., Ehira, S., Tanaka, K., Asai, K., and Ohta, N. (2008a) Molecular phylogeny and evolution of the plastid and nuclear encoded *cbbX* genes in the unicellular red alga *Cyanidioschyzon merolae*. *Genes Genet. Syst.*, *83*, 127-133.

Fujita, K., Tanaka, K., Sadaie, Y., and Ohta, N. (2008b) Functional analysis of the plastid and nuclear encoded CbbX proteins of *Cyanidioschyzon merolae. Genes Genet. Syst., 83*, 135-142.

Fujiwara, M., Nagashima, A., Kanamaru, K., Tanaka, K., and Takahashi, H. (2000) Three new nuclear genes, *sigD, sigE, sigF*, encoding putative plastid RNA polymerase sigma factors in *Arabidopsis thaliana. FEBS Lett., 485*, 178-182.

Galperin, M. Y. (2006) Structural classification of bacterial response regulators: diversity of output domains and domain combinations. *J. Bacteriol., 188*, 4169-4182.

Gibson, J. L., and Tabita, F. R. (1997) Analysis of the *cbbXYZ* operon in *Rhodobacter sphaeroides*. *J. Bacteriol.*, *179*, 663-669.

Glöckner, G., Rosenthal, A., and Valentin, K. (2000) The structure and gene repertoire of an ancient red algal plastid genome. *J. Mol. Evol., 51*, 382-390.

Hajdukiewicz, P. T., Allison, L. A., and Maliga, P. (1997) The two RNA polymerases encoded by the nuclear and the plastid compartments transcribed distinct groups of genes in tobacco plastids. *EMBO J.*, 16, 4041-4048.

Hanaoka, M., and Tanaka, K. (2008) Dynamics of RpaB-promoter interaction during high light stress, revealed by chromatin immunoprecipitation (ChIP) analysis in *Synechococcus elongatus* PCC 7942. *Plant J., 56*, 327-335.

Hedtke, B., Boerner, T., and Weihe, A. (1997) Mitochondrial and chloroplast phage-type RNA polymerases in *Arabidopsis. Science, 277*, 809-811.

Hsiao, H. Y., He, Q., van Waasbergen, L. G., and Grossman, A. R. (2004) Control of photosynthetic and high-light-responsive genes by the histidine kinase DspA: negative and positive regulation and interaction between signal transduction pathways. *J. Bacteriol., 186*, 3882-3888.

Ichikawa, K., Shimizu, A., Okada, R., Satbhai, S. B., and Aoki, S. (2008) The plastid sigma factor SIG5 is involved in the diurnal regulation of the chloroplast gene *psbD* in the moss *Physcomitrella patens. FEBS Lett., 582*, 405-409.

Imamura, S., Terashita, M., Ohnuma, M., Maruyama, S., Minoda, A., Weber, A. P., Inouye, T., Sekine, Y., Fujita, Y., Omata, T., and Tanaka, K. (2010) Nitrate assimilatory genes and their transcriptional regulation in a unicellular red alga *Cyanidioschyzon merolae*: genetic evidence for nitrite reduction by a sulfite reductase-like enzyme. *Plant Cell Physiol., 51*, 707-717.

Imamura, S., Kanesaki, Y., Ohnuma, M., Inouye, T., Sekine, Y., Fujiwara, T., Kuroiwa, H., Kuroiwa, T., and Tanaka, K. (2009) R2R3-type MYB transcription factor, CmMYB1, is a central nitrogen regulator in *Cyanidioschyzon merolae. Proc. Natl. Acad. Sci. USA, 106*, 12548-12553.

Iwasaki, H., William, S. B., Kitayama, Y., Ishiura, M., Golden, S. S., and Kondo, T. (2000) A kaiC-interacting sensory histidine kinase, SasA, necessary to sustain robust circadian oscillation in cyanobacteria. *Cell*, 101, 223-233.

Jacobs, M. A., Connell, L., and Cattolico, R. A. (1999) A conserved His-Asp signal response regulator-like gene in *Heterosigma akashiwo* chloroplasts. *Plant Mol. Biol., 41*, 645-655.

Kabeya, Y., Kobayashi, Y., Suzuki, H., Itoh, J., and Sugita, M. (2007) Transcription of plastid genes is modulated by two nuclear-encoded alpha subunits of plastid RNA polymerase in the moss *Physcometrella patens. Plant J.*, 52, 730-741.

Kanamaru, K., and Tanaka, K. (2004) Roles of chloroplast RNA polymerase sigma factors in chloroplast development and stress responses in higher plants. *Biosci. Biotechnol. Biochem., 68*, 2215-2223.

Kaneko, T., Sato, S., Kotani, H., Tanaka, A., Asamizu, E. et al. (1996) Sequence analysis of the genome of the unicellular cyanobacterium *Synechocystis* sp. strain PCC 6803. II. Sequence determination of the entire genome and assignment of potential protein-coding regions. *DNA Res., 3*, 109-136.

Kanesaki, Y., Yamamoto, H., Paithoonrangsarid, K., Shoumskaya, M., Suzuki, I., Hayashi, H., and Murata, N. (2007) Histidine kinases play important roles in the perception and signal transduction of hydrogen peroxide in the cyanobacterium, *Synechocystis* sp. PCC 6803. *Plant J., 49*, 313-324.

Kappel, A. D., and van Waasbergen, L. G. (2007) The response regulator RpaB binds the high light regulatory 1 sequence upstream of the high-light-inducible *hliB* gene from the cyanobacterium *Synechocystis* PCC 6803. *Arch. Microbiol., 187*, 337-342.

Kappel, A. D., Bhaya, D., and van Waasbergen, L. G. (2006) Negative control of the high light-inducible *hliA* gene and implications for the activities of the NblS sensor kinase in the cyanobacterium *Synechococcus elongatus* strain PCC 7942. *Arch. Microbiol., 186*, 403-413.

Kawakami, T., Sakaguchi, K., Takechi, K., Takano, H., and Takio, S. (2009) Ammonium induced expression of the red algal chloroplast gene Ycf18, a putative homolog of the cyanobacterial NblA gene involved in nitrogen deficiency-induced phycobilisome degradation. *Biosci. Biotechnol. Biochem., 73*, 740-743.

Lapidot, M., Raveh, D., Sivan, A., Arad, S. M., and Shapira, M. (2002) Stable chloroplast transformation of the unicellular red alga *Porphyridium species. Plant Physiol., 129*, 7-12.

Lemieux, C., Otis, C., and Turmel, M. (2007) A clade uniting the green algae *Mesostigma viride* and *Chlorokybus atmophyticus* represents the deepest branch of the Streptophyta in chloroplast genome-based phylogenies. *BMC Biol., 5*, 2.

Luque, I. Flores, E., and Herrero, A. (1994) Molecular mechanism for the operation of nitrogen control in cyanobacteria. *EMBO J., 13*, 2862-2869.

Maddocks, S. E., and Oyston, P. C. (2008) Structure and function of the LysR-type transcriptional regulator (LTTR) family proteins. *Microbiology, 154*, 3609-3623.

Maheswaran, M., Urbanke, C., and Forchhammer, K. (2004) Complex formation and catalytic activation by the PII signaling protein of N-acetyl-L-glutamate kinase from *Synechococcus elongatus* strain PCC 7942. *J. Biol. Chem., 279*, 55202-55210.

Maier, U. G., Fraunholz, M., Zauner, S., Penny, S., and Douglas, S. (2000) A nucleomorph-encoded CbbX and the phylogeny of RuBisCo regulators. *Mol. Biol. Evol.*, *17*, 576-583.

Marin, K., Suzuki, I., Yamaguchi, K., Ribbeck, K., Yamamoto, H., Kanesaki, Y., Hagemann, M., and Murata, N. (2003) Identification of histidine kinases that act as sensors in the perception of salt stress in *Synechocystis* sp. PCC 6803. *Proc. Natl. Acad. Sci. USA, 100*, 9061-9066.

Marin, B., Nowack, E. C. M., and Melkonian, M. (2005) A plastid in the making: Evidence for a second primary endosymbiosis. *Protist, 156*, 425-432.

Matsuzaki, M., Misumi, O., Shin-I, T., Maruyama, S., Takahara, M., et al. (2004) Genome sequence of the ultrasmall unicellular red alga *Cyanidioschyzon merolae* 10D. *Nature*, 428, 653-657.

Mikami, K., Kanesaki, Y., Suzuki, I., and Murata, N. (2002) The histidine kinase Hik33 perceives osmotic stress and cold stress in *Synechocystis* sp. PCC 6803. *Mol. Microbiol., 46*, 905-915.

Minoda, A., Weber, A. P. M., Tanaka, K., and Miyagishima, S. (2010) Nucleus-independent control of the Rubisco operon by the plastid-encoded transcription factor Ycf30 in the red alga *Cyanidioschyzon merolae*. *Plant Physiol.* (in press).

Minoda, A., Nagasawa, K., Hanaoka, M., Horiuchi, M., Takahashi, H., and Tanaka, K. (2005) Microarray profiling of plastid gene expression in a unicellular red alga, *Cyanidioschyzon merolae*. *Plant Mol. Biol., 59*, 375-385.

Mizuno, T., Kaneko, T., and Tabata, S. (1996) Compilation of all genes encoding bacterial two-component signal transducers in the genome of the cyanobacterium, *Synechocystis* sp. strain PCC 6803. *DNA Res., 3*, 407-414.

Nagashima, A., Hanaoka, M., Shikanai, T., Fujiwara, M., Kanamaru, K., Takahashi, H., and Tanaka, K. (2004) The multiple-stress responsive plastid sigma factor, SIG5, directs activation of the *psbD* blue light-responsive promoter (BLRP) in *Arabidopsis thaliana. Plant Cell Physiol., 45*, 357-368.

Nakayama, T., and Ishida, K. (2009) Another acquisition of a photosynthetic organelle is underway in *Paulinella chromatophora*. *Curr. Biol.*, 19, R284-285.

Nishimura, T., Takahashi, Y., Yamaguchi, O., Suzuki, H., Maeda, S., and Omata, T. (2008) Mechanism of low CO_2-induced activation of the *cmp* bicarbonate transporter operon by a LysR family protein in the cyanobacterium *Synechococcus elongatus* strain PCC 7942. *Mol. Microbiol., 68*, 98-109.

Nowack, E. C., Melkonian, M., and Gloeckner, G. (2008) Chromatophore genome sequence of *Paulinella* sheds light on acquisition of photosynthesis by eukaryotes. *Curr. Biol., 18*, 410-418.

Ohta, N., Matsuzaki, M., Misumi, O., Miyagishima, S.-Y., Nozaki, H., Tanaka, K., Shin-I, T., Kohara, Y., and Kuroiwa, T. (2003) Complete sequence and analysis of the plastid genome of the unicellular red alga *Cyanidioschyzon merolae*. *DNA Res., 10*, 67-77.

Oikawa, K., Tanaka, K., and Takahashi, H. (1998) Two types of differentially photo-regulated nuclear genes that encodes sigma factors for chloroplast RNA polymerase in the red alga *Cyanidium caldarium* strain RK-1. *Gene, 210*, 277-285.

Osanai, T., Ikeuchi, M., and Tanaka, K. (2008) Group 2 sigma factors in cyanobacteria. *Physiol. Plant., 133*, 490-506.

Osanai, T., and Tanaka, K. (2007) Keeping in touch with PII: PII-interacting proteins in unicellular cyanobacteria. *Plant Cell Physiol., 48*, 908-914.

Oudot-Le Secq, M.-P., Grimwood, J., Shapiro, H., Armbrust, E. V., Bowler, C., and Green, B. R. (2007) Chloroplast genomes of the diatoms *Phaeodactylum tricornutum* and *Thalassiosira pseudonana*: comparison with other plastid genomes of the red lineages. *Mol. Genet. Genomics, 277*, 427-439.

Paithoonrangsarid, K., Shoumskaya, M. A., Kanesaki, Y., Satoh, S., Tabata, S., Los, D. A., Zinchenko, V. V., Hayashi, H., Tanticharoen, M., Suzuki, I., and Murata, N. (2004) Five histidine kinases perceive osmotic stress and regulate distinct sets of genes in *Synechocystis*. *J. Biol. Chem., 279*, 53078-53086.

Puthiyaveetil, S., and Allen, J. F. (2009) Chloroplast two-component systems: evolution of the link between photosynthesis and gene expression. *Proc. R. Soc. B, 276*, 2133-2145.

Puthiyaveetil, S., Kavanagh, T. A., Cain, P., Sullivan, J. A., Newell, C. A., Gray, J. C., Robinson, C., van der Giezen, M., Rogers, M. B., and Allen, J. F. (2008) The ancestral symbiont sensor kinase CSK links photosynthesis with gene expression in chloroplasts. *Proc. Natl. Acad. Sci. USA, 105*, 10061-10066.

Reith, M., and Munholland, J. (1993) A high-resolution gene map of the chloroplast genome of the red alga *Porphyra purpurea*. *Plant Cell, 5*, 465-475.

Salinas, P., Ruiz, D., Cantos, R., Lopez-Redondo, M. L., Marina, A., and Contreras, A. (2007) The regulatory factor SipA provides a link between NblS and NblR signal transduction pathways in the cyanobacterium *Synechococcus* sp. PCC 7942. *Mol. Microbiol., 66*, 1607-1619.

Schell, M. A. (1993) Molecular biology of the LysR family of transcriptional regulators. *Annu. Rev. Microbiol., 47*, 597-626.

Schröter, Y. Steiner, S., Matthäi, K., and Pfannschmidt, T. (2010) Analysis of oligomeric protein complexes in the chloroplast sub-proteome of nucleic acid-binding proteins from mustard reveals potential redox regulators of plastid gene expression. *Proteomics, 10*, 2191-2204.

Seino, Y., Takahashi, T., and Hihara, Y. (2009) The response regulator RpaB binds to the upstream element of photosystem I genes to work for positive regulation under low-light conditions in *Synechocystis* sp. Strain PCC 6803. *J. Bacteriol., 191*, 1581-1586.

Seki, A., Hanaoka, M., Akimoto, Y., Masuda, S., Iwasaki, H., and Tanaka, K. (2007) Induction of a group 2 sigma factor, RPOD3, by high light and the underlying mechanism in *Synechococcus elongatus* PCC 7942. *J. Biol. Chem., 282*, 36887-36894.

Shoumskaya, M. A., Paithoonrangsarid, K., Kanesaki, Y., Los, D. A., Zinchenko, V. V., Tanticharoen, M., Suzuki, I., and Murata, N. (2005) Identical Hik-Rre systems are involved in perception, *J. Biol. Chem., 280*, 21531-21538.

Stragier, P., Richaud, F., Borne, F., and Patte, J. C. (1983) Regulation of diaminopimelate decarboxylase synthesis in *Escherichia coli*. I. Identification of a *lysR* gene encoding an activator of the *lysA* gene. *J. Mol. Biol., 168*, 307-320.

Sugiyama, K., Hayakawa, T., Kudo, T., Ito, T., and Yamaya, T. (2004) Interaction of N-acetylglutamate kinase with PII-like protein in rice. *Plant Cell Physiol., 45*, 1768-1778.

Suzuki, I., Kanesaki, Y., Mikami, K., Kanehisa, M., and Murata, N. (2001) Cold-regulated genes under control of the cold sensor Hik33 in *Synechocystis*. *Mol. Microbiol., 40*, 235-244.

Suzuki, I., Los, D. A., Kanesaki, Y., Mikami, K., and Murata, N. (2000) The pathway for perception and transduction of low-temperature signals in *Synechocystis*. *EMBO J., 19*, 1327-1334.

Takai, N., Nakajima, M., Oyama, T., Kito, R., Sugita, C., Sugita, M., Kondo, T., and Iwasaki, H. (2006) A KaiC-associated SasA-RpaA two-component regulatory system as a major circadian timing mediator in cyanobacteria. *Proc. Natl. Acad. Sci. USA, 103*, 12109-12114.

Tanaka, K., Tozawa, Y., Mochizuki, N., Shinozaki, K., Nagatani, A., Wakasa, K., and Takahashi, H. (1997) Characterization of three cDNA species encoding plastid RNA polymerase sigma factors in *Arabidopsis thaliana*: evidence for the sigma factor heterogeneity in higher plant plastids. *FEBS Lett., 413*, 309-313.

Tanaka, K., Oikawa, K., Ohta, N., Kuroiwa, H., Kuroiwa, T., and Takahashi, H. (1996) Nuclear encoding of a chloroplast RNA polymerase sigma subunit in a red alga. *Science, 272*, 1932-1935.

Tanigawa, R., Shirokane, M., Maeda, S.-I., Omata, T., Tanaka, K., and Takahashi, H. (2002) Transcriptional activation of NtcA-dependent promoters of *Synechococcus* sp. PCC 7942 by 2-oxoglutarate *in vitro*. *Proc. Natl. Acad. Sci. USA, 99*, 4251-4255.

Terashita, M., Maruyama, S., and Tanaka, K. (2006) Cytoplasmic localization of the single glutamine synthetase in a unicellular red alga, *Cyanidioschyzon merolae* 10D. *Biosci. Biotechnol. Biochem., 70*, 2313-2315.

Tsunoyama, Y., Ishizaki, Y., Morikawa, K., Kobori, M., Nakahira, Y., Takeba, G., Toyoshima, Y., and Shiina, T. (2004) Blue light-induced transcription of plastid-encoded *psbD* gene is mediated by a nuclear-encoded transcription initiation factor, AtSig5. *Proc. Natl. Acad. Sci. USA, 101*, 3304-3309.

Tu, C. J., Shrager, J., Burnap, R. L., Postier, B. L., and Grossman, A. R. (2004) Consequences of a deletion in *dspA* on transcript accumulation in *Synechocystis* sp. strain PCC 6803. *J. Bacteriol., 186*, 3889-3902.

van Keulen, G., Girbal, L., van den Bergh, E. R., Dijkhuizen, L., and Meijer, W. G. (1998) The LysR-type transcriptional regulator CbbR controlling autotrophic CO_2 fixation by *Xanthobacter flavus* is an NADPH sensor. *J. Bacteriol., 180*, 1411-1417.

van Waasbergen, L. G., Dolganov, N., and Grossman, A. R. (2002) *nblS*, a gene involved in controlling photosynthesis-related gene expression during high light and nutrient stress in *Synechococcus elongatus* PCC 7942. *J. Bacteriol., 184*, 2481-2490.

Vazquez-Bermudez, M. F., Herrero, A., and Flores, A. (2002) 2-Oxoglutarate increases the binding affinity of the NtcA (nitrogen control) transcription factor for the *Synechococcus glnA* promoter. *FEBS Lett., 512*, 71-74.

Vega-Palas, M. A., Flores, E., and Herrero, A. (1992) NtcA, a global nitrogen regulator from the cyanobacterium *Synechococcus* that belongs to the Crp family of bacterial regulators. *Mol. Microbiol., 6*, 1853-1859.

Vidal, R., Lopez-Maury, L., Guerrero, M. G., and Florencio, F. J. (2009) Characterization of an alcohol dehydrogenase from the cyanobacterium *Synechocystis* sp. strain PCC 6803 that responds to environmental stress conditions via the Hik34-Rre1 two-component system. *J. Bacteriol., 191*, 4383-4391.

Weber, P., Fulgosi, H., Piven, I., Muller, L., Krupinska, K., Duong, V. H., Herrmann, R. G., and Sokolenko, A. (2006) TCP34, a nuclear-encoded response regulator-like TPR protein of higher plant chloroplasts. *J. Mol. Biol., 357*, 535-549.

Yoon, H. S., Nakayama, T., Reyes-Prieto, A., Andersen, R. A., Boo, S. M., Ishida, K., and Bhattacharya, D. (2009) A single origin of the photosynthetic organelle in different *Paulinella* lineages. *BMC Evol. Biol., 9*, 98.

Yoon, H. S., Reyes-Prieto, A., Melkonian, M., and Bhattacharya, D. (2006) Minimal plastid genome evolution in the *Paulinella* endosymbiont. *Curr. Biol., 16*, R670-672.

Yoshimura, H., Hisabori, T., Yanagisawa, S., and Ohmori, M. (2000) Identification and characterization of a novel cAMP receptor protein in the cyanobacterium *Synechocystis* sp. PCC 6803. *J. Biol. Chem., 275*, 6241-6245.

Zhao, M.-X., Jiang, Y.-L., He, Y.-X., Chen, Y.-F., Teng, Y.-B., Chen, Y., Zhang, C.-C., and Zhou, C.-Z. (2010) Structural basis for the allosteric control of the global transcription factor NtcA by the nitrogen starvation signal 2-oxoglutarate. *Proc. Natl. Acad. Sci. USA* (published online).

In: Porphyra yezoensis
Editor: Koji Mikami
ISBN 978-1-61122-304-0

Chapter 4

RETROTRANSPOSONS IN *PORPHYRA YEZOENSIS*

Susumu Takio
Center for Marine Environment Studies, Kumamoto University,
Kurokami 2-39-1, Kumamoto 860-8555, Japan

ABSTRACT

Retrotransposons are mobile genetic elements that transpose via an RNA intermediate. They are extremely abundant in higher plants and play a major role in plant gene and genome evolution. They can be divided into two broad groups: the long terminal repeat (LTR) and non-LTR retrotransposons. Both types have provided valuable tools for basic and applied studies in plant biology. From comprehensive whole-genome analysis, information about retrotransposons in algae, including green and brown algae, is increasing, while less is known about red algal retroelements. This chapter describes the gene structures of LTR and non-LTR retrotransposons recently isolated from *Porphyra yezoensis* and discusses them from the perspective of the evolution of retrotransposons in plants.

1. INTRODUCTION

Retrotransposable elements, which transpose through the reverse transcription of an mRNA intermediate, are the most abundant and widespread class of eukaryotic transposable elements (Kumar and Bennetzen, 1999). Most retrotransposons are defective elements that cannot transpose. Because these defective elements are fixed at specific sites in chromosomes, they can be used as molecular markers when they are abundant in the genome. In higher plants, some active retrotransposons have provided potential tools for efficient mutagen and genetic analysis using gene tagging (Kumar and Hirochika, 2001; Mazier et al., 2007)

Recent comprehensive analysis of entire genomes in algae has shown that major retrotransposons are present not only in seed plants, but also in unicellular algae, although

Author for correspondence: e-mail stakio@gpo.kumamoto-u.ac.jp

algae contain fewer and less variable retrotransposons than seed plants. Unlike unicellular green algae (Derelle et al., 2006; Marchant et al., 2007), diatoms (Armbrust et al., 2004; Bowler et al., 2008) and cryptomonad algae (Khan et al., 2007), the unicellular red alga *Cyanidioschyzon merolae* genome (Matsuaki et al., 2004) lacks typical retrotransposons (Nozaki et al., 2007). Because red algae are one of the most primitive lineages in plant evolution, retrotransposons in red macroalgae are important in the evolution of retrotransposons in plants. This chapter describes the structural properties of retrotransposons identified from *Porphyra yezoensis* and discusses them from the perspective of the evolution of retrotransposons in plants.

2. Classification of Retrotransposons in Plants

Retrotransposons can be divided into two broad groups: the long terminal repeat (LTR)-containing retrotransposons and non-LTR retrotransposons (Figure 1) (Kumar and Bennetzen. 1999). LTR retrotransposons are further classified into two families: the Tyl/*copia* and Ty3/*gypsy* families (referred to as *copia* and *gypsy* in this chapter; Figure 1A). *Copia* and *gypsy* retrotransposons encode all of the proteins required for their transposition, including a viral particle coat (*Gag*), protease (Prot), integrase (IN), reverse transcriptase (RT), and RNase H (RH), which converts mRNA into cDNA and integrates it into the genome, thereby increasing the copy number of the element. The two types can be distinguished by nucleotide sequence differences in their catalytic enzymes and the inverse order of the integrase and RT/RNase H genes.

Non-LTRs include long-interspersed elements (LINEs) and short-interspersed elements (SINEs). LINEs usually have two separate open-reading frames (ORFs): ORF1 encodes a *gag*-like protein and ORF2 encodes the endonuclease and reverse transcriptase required for self-propagation (Schmidt, 1999). Thus, LTR retrotransposons and LINEs are also called autonomous retrotransposons. Unlike these autonomous elements, SINEs lack any discernible ORF (non-autonomous elements). The mechanism of the transposition of non-autonomous elements remains unclear, and it has been suggested that the reverse transcriptase responsible for the retrotransposition of SINEs is provided in *trans* by some LINEs because the 3'-ends of several families of SINEs are homologous with the 3'-ends of some LINEs (Okada et al., 1997). Among algal retrotransposons, the SINE (SINE6-CR) and LINE (RandI-1) in *Chlamydomonas reinhardtii* have conserved sequences (Rebpase, Kohany et al., 2006).

3. Copia-Like Elements from *Porphyra Yezoensis*

Copia and *gypsy* retrotransposons do not show nucleotide sequence homology. Nevertheless, a pair of degenerate oligonucleotide primers based on two highly conserved domains of RT can be used for PCR amplification of *copia*-like and *gypsy*-like sequences. Using degenerate primer sets for *copia*- and *gypsy*-like elements, several DNA fragments encoding the RT related to retrotransposons were isolated from the gametophyte of *P. yezoensis* (Figure 2A) (Zhang et al., 2006). In a public database of *P. yezoensis* expressed sequence tags (ESTs; Kazusa DNA Research Institute in Japan; Asamizu et al., 2003; Nikaido et al., 2000), most of the EST clones annotated as retrotransposon-related sequences are *gypsy*-like elements. AV433981 is one of these clones. In contrast, AV437032 is a *copia*-like

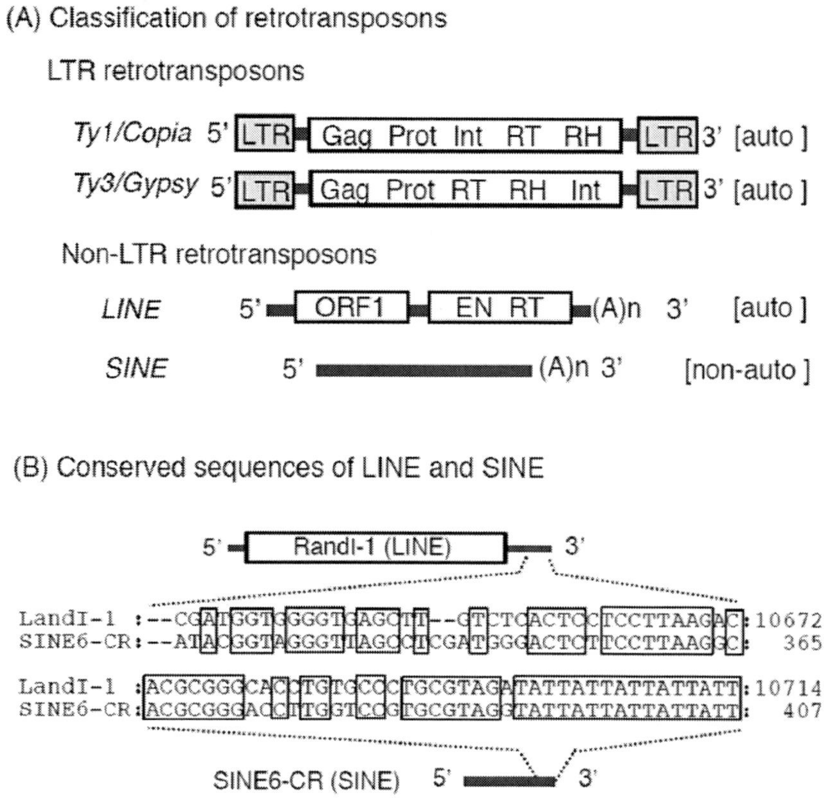

Figure 1. Classification of retrotransposons. (A) A schematic representation of the gene structure of typical retrotransposons. The ORF and LTR are indicated by white and gray boxes, respectively. The ORF of LTR retrotransposons encodes the gag protein (Gag), protease (Prot), integrase (IN), reverse transcriptase (RT), and RNase H (RH). The second ORF of LINE encodes endonuclease (EN) and RT. [auto] and [non-auto] at the right of the element suggest autonomous and non-autonomous elements, respectively. (B) Conserved sequences of the LINE (RandI-1) and SINE (SINE6-CR) in *Chlamydomonas reinhardtii.*

Element (Figure 2B). Genomic Southern analysis suggested that *copia*-like elements have fewer copy numbers in the genome than *gypsy*-like elements.

From these *copia*-like gene fragments, three *copia*-like RT-containing genes, named *PyRE10G* (Peddigari et al., 2008a), *PyRE1G1* (Peddigari et al., 2008b), and *PyRE2A* (Zhang et al., 2006), have been isolated (Figure 2C). *PyRE1G1* has the typical gene structure of *copia* elements, including an ORF encoding *gag*, proteinase, integrase, RT, and RNase H, in that order, and LTRs at both ends of the gene. *PyRE10G* has a putative ORF encoding *gag*, proteinase, integrase, RT, and RNase H with one-stop codon in the frame, and lacks LTR sequences at both ends. Unlike typical retrotransposons, *PyRE2A* is a small gene, about 2.8 kb in size, and encodes a single RT/RNase H gene. Because the putative ORFs of *PyRE10G* and

PyRE2A contain stop codon(s) in-frame, they are defective elements, while *PyRE1G1* has the gene structure typical of *copia* retrotransposons.

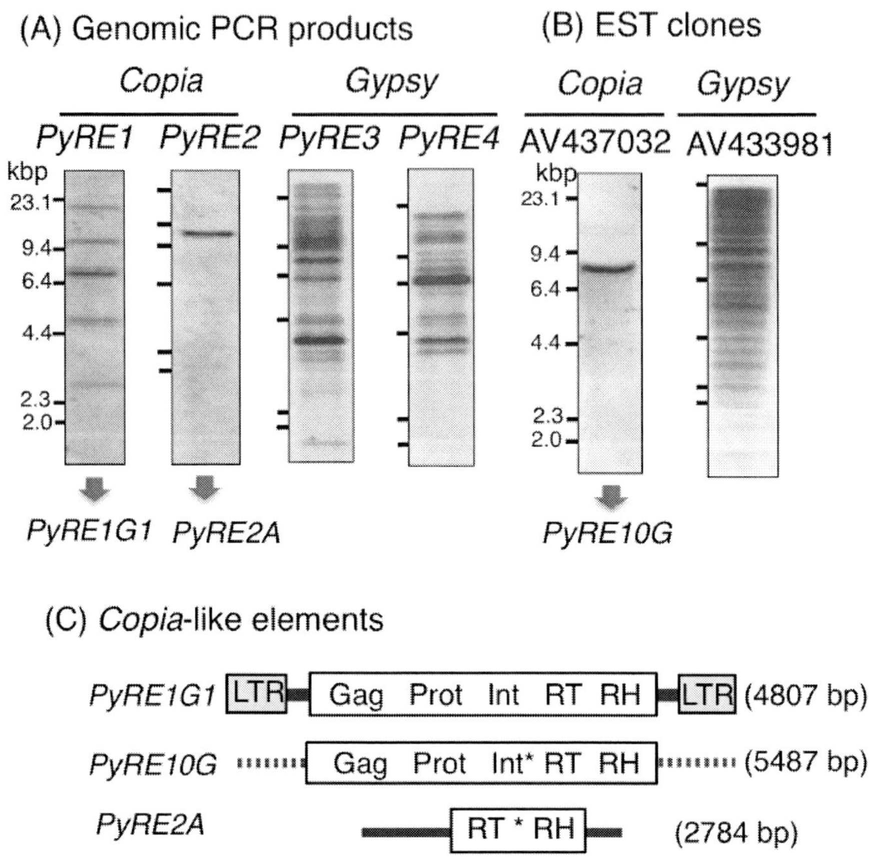

Figure 2. Isolation of *copia*-like elements from *Porphyra yezoensis*.(A) Southern blot analysis using a probe derived from the genomic PCR products. An approximately 270-bp RT fragment of *PyRE1* and *PyRE2* was isolated by genomic PCR using a degenerate primer designed for the RT of *copia* retrotransposon. An approximately 270-bp RT fragment of *PyRE3* and *PyRE4* was isolated using the primer designed for *gypsy* retrotransposons. The full-length genes of *PyRE1G1* and *PyRE2A* were isolated from *PyRE1* and *PyRE2*, respectively. (B) Southern blot analysis using a probe derived from EST clones of *P. yezoensis*. AV437032 and AV433981 were most closely related to *copia* and *gypsy* retrotransposons, respectively. The full-length *PyRE10G* gene was isolated from AV437032. (C) The gene structure of the full-length genes from *P. yezoensis*.

In addition to the coding region, the integrity of the untranslated region (UTR), including the LTR, primer binding site (PBS), and polypurine tract (PPT) sequences are also important for replication. Figure 3 compares the structure of the LTR and PBS sequences of *copia* retrotransposons in seed plants and some algae. Although the size of the LTR differs among species, the 5'-LTR sequences are completely identical to the 3'-sequence in most elements except *PyRE1G1*, *AtRE1*, and *Copia*. In *Copia* and *AtRE1*, a single nucleotide is deleted in either LTR, but other sequences in the LTR are identical to that of other LTR. *PyRE1G1* contains 204-bp LTRs at both ends. The nucleotide sequences of the 5'-LTR are identical to those of the 3'-LTR, except for a single nucleotide difference. Both LTRs have identical 5'-TG...CA-3' dinucleotide end sequences to other retrotransposons.

In many retrotransposons, PBS is complementary to the 3'-terminal portion of the cellular initiator methionine tRNA, tRNA(iMet).

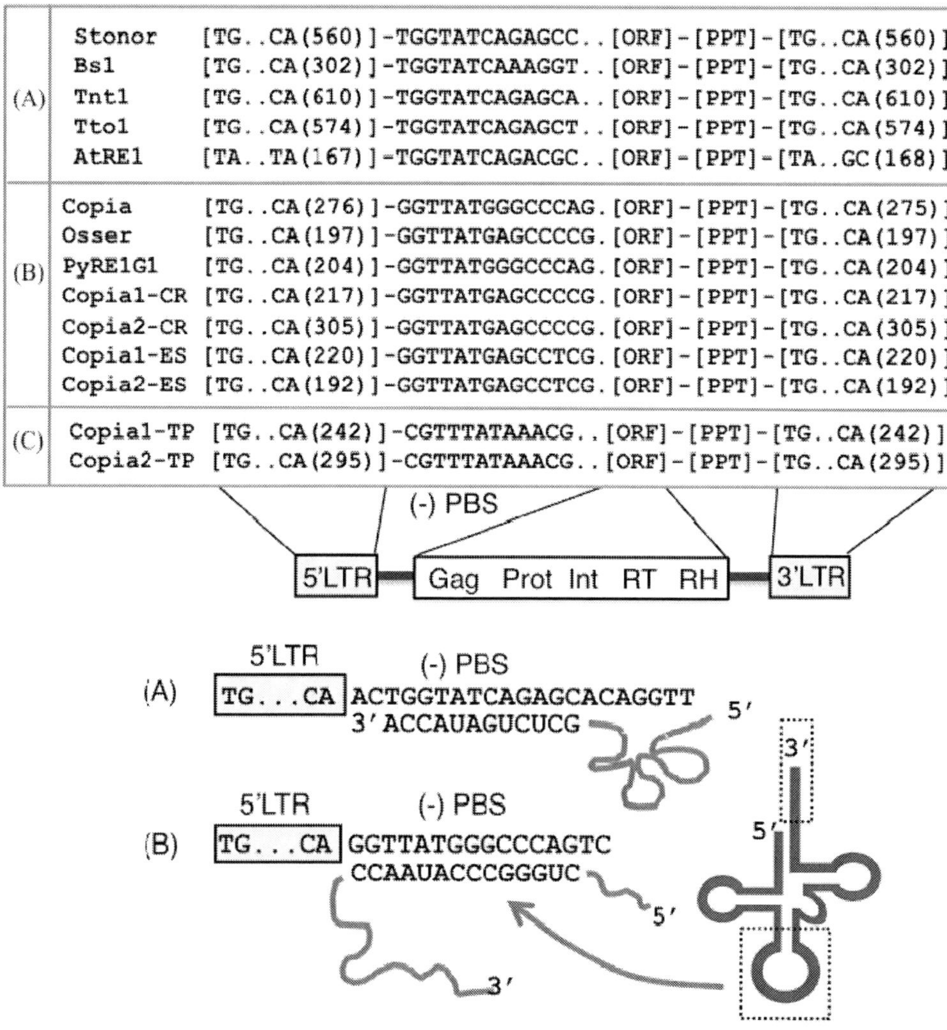

Figure 3. Comparison of the LTR and (–) PBS sequences of *PyRE1G1* and those of other *copia* retrotransposons. (A) *Copia* retrotransposons in which PBS is complementary to the 3'-terminal portion of cellular tRNA(iMet). (B) *Copia* retrotransposons in which PBS is complementary to the internal portion of cellular tRNA(iMet). (C) *Copia* retrotransposons in which PBS includes palindrome sequences. The figure in parentheses in LTR shows the nucleotide number in LTR.

This is common in higher plant *copia* retrotransposons, such as those of maize *Stonor* and *Bs1*, tobacco *Ttn1* and *Tto1*, and *Arabidopsis AtRE1* (Figure 3A). In contrast, the *Drosophila copia* uses an internal portion of tRNA(iMet) as a primer for minus strand DNA synthesis. Similar PBS sequences are found in *PyRE1G1*. Further, they are also present in *Volvox Osser*, *Chlamydomonas Copia1-CR* and *Copia2-CR*, and the brown alga *Ectocarpus Copia1-ES* and *Copia2-ES* (Figure 3B). The preferential use of the internal portion of tRNA(iMet) as a PBS among algae suggests that this type of PBS may be the ancestral mode of primer binding. By comparison, diatom *Thalassiosira copia* retrotransposons do not possess a tRNA-like PBS; rather, they seem to use self-priming with 12-bp palindrome sequences in *Copia1-TP* and *Copia2-TP* at the 3'-flanking region of the 5'-LTR (Figure 3C). These findings suggest that these *Thalassiosira copia* retrotransposons are unrelated to the red algal *PyRE1G1* and *copia* elements of green and brown algae.

In many copia retrotransposons, a polypurine tract (PPT) for plus-strand DNA synthesis (5'-GTGGGGGAG-3') is present upstream from the 3'-LTR; examples include 5'-GAGGGGGCG-3' in *copia* and 5'-GCGTGGGAG-3' in *Volvox Osser*. In *PyRE1G1*, similar sequences (5'-GTGGGGGAG-3') are present upstream from the 3'-LTR.

Based on the sequences of the ORF, LTRs, PBS, and PPT, *PyRE1G1* seems to be an intact retrotransposon capable of self-proliferation and transposition.

4. SINE-Like Retrotransposons in *P. yezoensis*

In seed plants, LINEs can be isolated using degenerate primers based on the conserved domains of RT. However, no LINE element was isolated from *P. yezoensis* using these primers. Instead, two SINE-like elements, named *PySN1* and *PySN2*, were identified in the 760-bp PCR product using these primers by chance (Zhang et al., 2007).

The first plant SINE, which was named *p-SINE1*, was identified in the introns of the *Waxy* gene of *Oryza sativa* (Umeda et al., 1991). Since then, many plant SINE families have been identified, including *p-SINE2-3* in rice, *TS* in tobacco, *S1* in Brassica, and *AtSN1* and *AtSN2* in *Arabidopsis* (Yoshioka et al., 1993; Deragon et al., 1994; Lenoir et al., 2001; Myouga et al., 2001; Xu et al., 2005; Fawcett et al., 2006). In contrast, only three SINE elements have been found in non-vascular plants: in the green algae *Chlorella* (Noutoshi et al., 1998), *Chlamydomonas* (SINE-6_CR in Repbase; Kohany et al., 2006), and *Porphyra PySNs*. Plant SINEs usually contain an internal promoter with motifs (A box and B box) recognized by RNA polymerase III (Galli et al., 1981) , and target site duplications (TSDs) at both ends (Figure 4A). *Porphyra PySN1* and *PySN2* contain these sequences. Based on the sequences of the 5'-head region, SINEs may have evolved from three different small RNAs: tRNA, 7SL RNA, and 5S rRNA. Plant SINEs are derived from tRNA. Thus, a degenerate cloverleaf tRNA secondary structure is predicted for the 5'-half of some plant SINEs: e.g., rice *p-SINE1-1*, *Brassica S1-na2*, and tobacco *TS-Nt2*. *Porphyra PySN1* and *PySN2* also have the degenerate tRNA secondary structure. These findings suggest that *PySN1* and *PySN2* are tRNA-derived SINEs, like plant SINEs. However, they do not show significant sequence homology with other plant SINEs, suggesting that *PySN1* and *PySN2* represent two novel SINE families.

PySN1 and *PySN2* are co-located within a 760-bp PCR fragment. To determine whether the two elements exist as partners throughout the genome, total DNA was digested with *Mva*I, which has one restriction site within each element and one site between the two elements. Each probe generated a single band (Figure 4B). The sizes of the bands (230 bp with the *PySN1* probe and 440 bp with the *PySN2* probe) were identical to the size predicted from the digestion of the 760-bp PCR fragment with *Mva*I. This suggests that the two elements exist as partners throughout the genome. Based on the haploid genome size of *P. yezoensis* (260 Mb;

Kaprun et al., 1991; Le Gall et al., 1993), the copy number of *PySN1* and *PySN2* in the genome was calculated to be 4,600 and 8,900, respectively.

Chlorella SINEs are abundant in the genome and show target site duplication at both ends in each gene family, but they lack the A-B box and cloverleaf tRNA secondary structure.

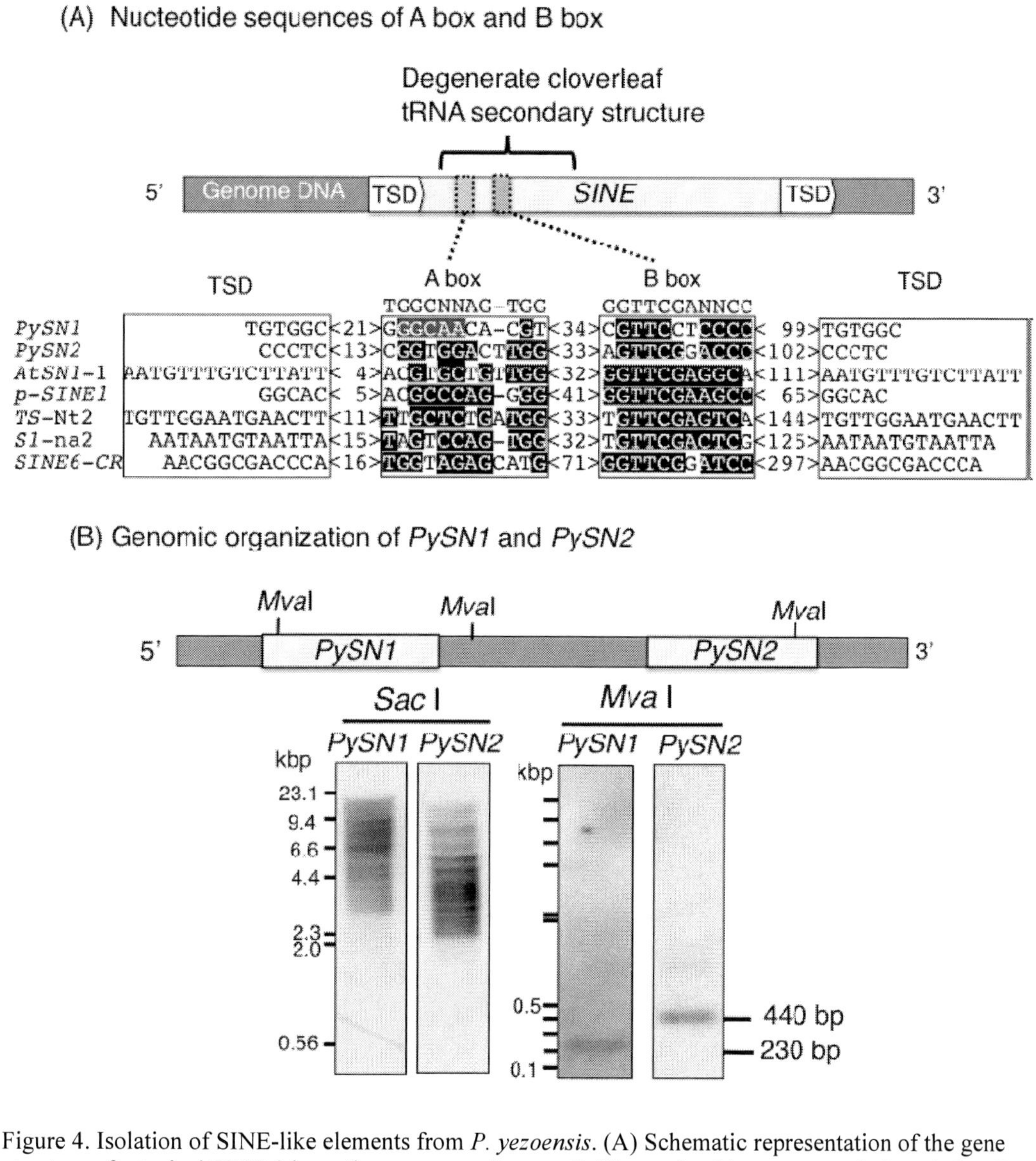

Figure 4. Isolation of SINE-like elements from *P. yezoensis*. (A) Schematic representation of the gene structure of a typical SINE (above the sequence alignment). The nucleotide sequences in box A, box B, and TSDs of *PySN1* and *PySN2* in *P. yezoensis*, *Arabidopsis AtSN1-1* (AB053466), rice *p-SINE1* (AB201734), tobacco *TS-Nt2* (D17454), Brassica *S1-na2* (AF101144), and *Chlamydomonas SINE6-CR* (Repbase, Kohany et al., 2006). (B) Genomic organization of *PySN1* and *PySN2*. Gene map of *PySN1* and *PySN2* (upper panel) and Southern blot analysis using gene-specific probes (lower panel). Genomic DNA was digested with the restriction enzyme indicated in the figure, and the digested DNA was transferred to a membrane and hybridized with the gene-specific probes.

Furthermore, they do not show any significant homology with the LINE elements found in the same organisms. Thus, the origin and nature of *Chlorella* SINEs remain to be determined. In contrast to *Chlorella* SINEs, the *Chlamydomonas* SINE *SINE6-CR* possesses typical structures, such as the A-B box and degenerate cloverleaf tRNA secondary structure. Furthermore, the 3'-end sequences of *SINE6-CR* are homologous to that of the LINE element RandI-1 (Figure 1B). The RandI family genes consist of six subfamilies and a total of 370 genes. All six gene subfamilies have the complete LINE structure, and each subfamily has 20 to 100 genes, suggesting that all of the subfamily genes are active. However, only the RandI-1 subfamily possesses 3'-end sequences homologous to that of *SNIE6-CR*. Of the 370 RandI family genes in the genome, 100 copies are RandI-1.These properties of the gene structure of RandI-1 suggest that RandI-1 helps *SINE6-CR* transposition. However, *SINE6-CR* exists in only 18 copies in the genome, which is much less than most SINEs. Considering the small number of SINEs in *Chlamydomonas*, the abundance of SINEs seems to be a unique feature of *Porphyra*.

5. Transcriptional Regulation of *Copia*-Like Elements and SINE-Like Elements

Although there are many copies in the host genome, their transcription and transposition are repressed under normal conditions. In higher plants, more than 20% of the cytosine in nuclear DNA is methylated, and DNA methylation is one mechanism used to silence the retrotransposons (Finnegan et al., 1998). Compared with higher plants, the green alga *Chlamydomonas* has much less methylated cytosine (Hattman et al., 1978) and the methylation status in many other algae remains unknown. Thus, the methylcytosine status of *Porphyra* retrotransposon *PyRE1G1* was analyzed, using a methyl-sensitive and methyl-insensitive isochizomer pair *Hpa*II/*Msp*I (Fukuda et al., 2004). No difference in restriction pattern between the pair was detected, suggesting that *PyRE1G1* is nearly free from cytosine methylation. In similar experiments, other retrotransposon-related genes also appeared to be free from cytosine methylation (unpublished data).

In higher plants, the transcription of retrotransposons is generally repressed under normal growth conditions. In *Porphyra*, three *copia*-like elements can be amplified by RT-PCR under normal growth conditions, but no signal is detected with Northern blotting, suggesting that the expression of these elements is repressed to a low level under normal growth conditions.

Although most plant retrotransposons seem to be inactive, ancient movable elements because of their defective structure, some active elements have also been identified. Of the active retrotransposons in plants, rice *Tos17* (Hirochika et al., 1996) and tobacco *Tnt1* (Grandbastien et al., 1989) and *Tto1* (Hirochika, 1993), which are *copia* retrotransposons, are the best characterized, and *Tos17* has provided valuable tools for genome analysis in rice as a endogenous insertional mutagen, and cloning of the tagged gene has been used in forward genetics to unravel gene functions (Hirochika, 2001; Kumar and Hirochika, 2001; Miyao et al., 2007).

The expression of these active elements is generally repressed under normal conditions and activated by biotic and abiotic stress factors, such as protoplast isolation, ultraviolet (UV) light, ozone, pathogens, wounding, and drought (Grandbastien, 1998). However, the responses of these retrotransposons differ among species. For example, *Tnt1* transcription was activated by protoplast isolation (Pouteau et al., 1991), but was not detectable in cell culture (Hirochika, 1993), while both stress factors activated the expression of *Tto1* and *Tos17* (Hirochika, 1993; Hirochika et al., 1996). Further, an unusual stress response has been reported for the *copia* retrotransposon *Reme1* in melon in which the expression of *Reme1*was activated by UV light, but not by wounding (Ramallo et al., 2008). Thus, to be able to apply plant retrotransposons, it is important to identify a stress factor that effectively activates the retrotransposon.

To determine whether the retrotransposons or related elements in *P. yezoensis* were transcriptionally active, the effects of various stress treatments on the accumulation of retrotransposon transcripts have been investigated, such as protoplast isolation, nutrient deficiency, and the application of methyl viologen, copper, and chitosan. Figure 5 shows the transcriptional response of *Porphyra copia*-like and *SINE*-like genes to protoplast treatment. *PyRE1G1* expression was activated after protoplast isolation, while *PyRE2A* and *PyRE10G* were not (Figure 5). Other stress treatments, such as the application of excess copper or chitosan, also activated the expression of *PyRE1G1*, but not *PyRE2A* or *PyRE10G* (unpublished data). Further, an underwater pulsed discharge treatment induced the transcriptional activation of *PyRE1G1*, but not *PyRE2A* or *PySNs* (Wang et al., 2010). *PyRE1G1* has a complete gene structure, as do retrotransposons with LTRs, while *PyRE2A* and *PyRE10G* lack this. Because the LTRs of the *Tnt1*, *Tto1*, and *OARE-1* retrotransposons contain stress-responsive *cis*-elements (Grandbastien, 1998; Grandbastien et al., 1997; Takeda et al., 1997), it is likely that the LTR of *PyRE1G1* governs the *cis*-element response to specific stress factors.

PySN1 and *PySN2* are predicted to exist as partners throughout the genome. However, they differ in transcript abundance: most *PySN2* family genes are repressed, while the *PySN1* family genes are expressed under normal conditions. Based on 3'- and 5'-rapid amplification of cDNA ends (RACE) experiments, *PySN1* genes are expressed as a chimera transcript with unrelated flanking SINE sequences and possess a poly(A) tail at the same position near the 3'-end of *PySN1*. The first step in the retrotransposition of SINEs is transcription by polymerase III. Because 5'-RACE products using the *PySN1* primers consisted of chimera transcripts with unrelated sequences, *PySN1* is not expressed as a single gene; instead, it appears to be used to provide polyadenylation signals at the 3'-end of the chimera transcript with unrelated sequences. This mode of *PySN1* transcription occurs under normal growth conditions. As shown in Figure 4B, the amount of *PySN1*-containing transcripts was increased after protoplast isolation, but all of the transcripts analyzed were chimeras.

Previously, SINEs were considered to be genomic parasites that conferred no benefit to the organism; however, they are now recognized as being beneficial to the host. For example, in animal SINEs, polyadenylation has been reported to generate functional proteins, such as the gamma subunit of muscle phosphorylase kinase (Maichele *et al.,* 1993) and leukaemia inhibitory factor (Michel et al., 1997). More recently, SINEs were postulated to be involved in regulating microRNA (miRNA) processing (Voinnet, 2009). miRNAs are a distinct class of small RNAs present in most eukaryotes. They arise from typical imperfect stem-loop structures, are processed by Dicer enzymes, and the mature 21-22-nt-long sequence is incorporated into a silencing complex, resulting in translational inhibition or mRNA degradation. Over-expression of SINE RNA in *Arabidopsis* generated phenotypes resembling those of miRNA-deficient mutants (Voinnet, 2009). It was found that the stem loops of SINE elements mimic the hairpin structure of miRNA precursors and bind to HYL1, the double stranded RNA-binding protein required for pri- to pre-miRNA processing (Pouch-Pélissier et al., 2008). These results suggest the tissue or developmental production of plant miRNA by competing for HYL1 and that SINE elements are able to regulate the miRNA processing pathway.

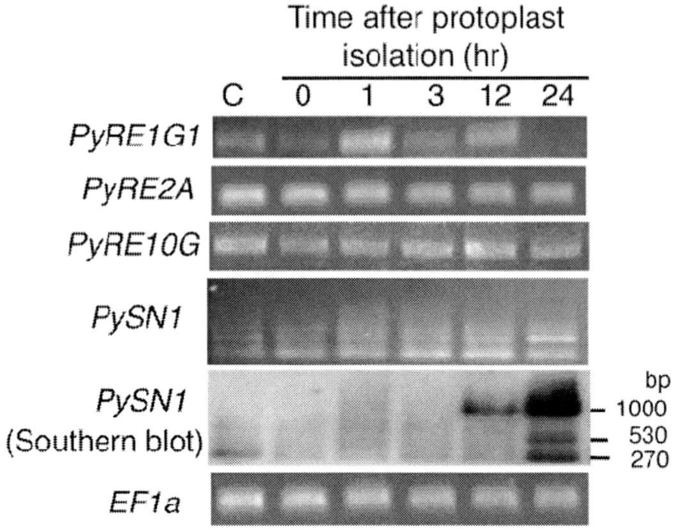

Figure 5. Changes in the transcript abundance of *copia*-like elements and *PySN1* after protoplast isolation. Protoplasts were prepared using papain for the pre-treatment of gametophytes and abalone acetone extracts for the digestion. RT-PCR was performed using the primer sets for *PyRE1G1*, *PyRE2A*, *PyRE10G*, and *EF1a*. The PCR products were electrophoresed in 1% agarose gels and stained with ethidium bromide (EtBr). With the *PySN1* primers, 5'-RACE was performed using the gene-specific primers. The PCR products were detected with EtBr staining and Southern blot analysis.

In *Porphyra*, *PySN1* family genes are transcriptionally active, and their expression is promoted by several stresses (Figure 5). To date, no sequences related to known proteins have been identified in the 5'-flanking region of the chimera transcripts. This suggests that *Porphyra* SINEs are used to generate functional proteins. Further, *PySN* transcripts may affect miRNA processing if *Porphyra* has miRNA regulation, as observed in higher plants. Whole genome analysis of *Porphyra* will provide valuable information about the relationship between SINEs and miRNA.

6. Phylogenetic Analysis of *Porphyra* Retrotransposons

In higher plant *copia* and *gypsy* retrotransposons, the protein motifs in *gag*, integrase, and RT have been identified by several motif programs. In *PyRE1G1*, the conserved motifs in *gag*, integrase, RT and RNase H were successfully predicted by the program HMMPfam (http://www.ebi.ac.uk/InterProScan/), including the CCHC motif in nucleic acid-binding *gag* protein (Mount and Rubin, 1985), the DNA-binding HHCC domain, and the catalytic DD(35)E domain conserved in integrases (Peterson-Burch and Boytas, 2002). The *PyRE1G1*

ORF and putative *PyRE10G* ORF encode the *gag*, integrase, and RT proteins containing these conserved domains. In the RT-like region of *copia* retrotransposons from various organisms, six conserved motifs are recognized (Xiong and Eickbush, 1990). PyRE1G1 and PyRE10G contain all six motifs (Peddigari et al., 2008a, 2008b). Based on the gene structure and protein motifs, *PyRE1G1* and *PyRE10G* can be classified as *copia* retrotransposons. This is supported by the phylogenetic tree constructed using the amino acid sequences of the RT region (Figure 6). Conversely, the phylogenetic tree suggests that PyRE1G1 is distinct from PyRE10G: PyRE1G1 is grouped in the clade of usual *copia* retrotransposons, while PyRE10G is in another clade, the GalEa group (Peddigari et al., 2008a, 2008b). Terrat et al. (2008) reported that GalEa elements isolated from galatheid crabs were closely related to *copia*-like elements in phylogenetically distant animals, such as urochordates (*Ciona intestinalis*), medaka (*Oryzias latipes*), and zebrafish (*Danio rerio*), and concluded that the GalEa group was a new clade of *copia*-like retrotransposons in aquatic animals (Terrat et al., 2008). Unlike *PyRE1G1*, *PyRE10G* is closely related to GalEa group elements. This was supported by phylogenetic trees constructed using integrase (Peddigari et al., 2008b) and RNase H (unpublished). Presently, however, LTR sequences have not been identified in *PyRE10G*, whereas most GalEa group elements do possess LTR sequences at both ends. Considering the distribution of GalEa elements among phylogenetically distant aquatic animals, Terrat et al. (2008) speculated that horizontal gene transfer was one reason for the unusual distribution. The inclusion of PyRE10G in the clade of GalEa elements indicates that GalEa group elements exist in plants. However, *GalEa*-like sequences are not present in higher plants and algae, except *Porphyra PyRE10G*. Because *GalEa* group elements exist in multicopy genes in the animal genome (Terrat et al., 2008), they seem to have been transposed as an active element. Because the distribution of *GalEa1* elements differs among different species in the same genus, it is thought that their transposition was not a very old event. By comparison, *PyRE10G* exists as a single copy in the genome, lacks the normal gene structure, has no LTR, and has an in-frame stop codon. Thus, *PyRE10G* might have become a defective element soon after invading the *Porphyra* genome. If one considers the horizontal transfer of *GalEa* elements to red algae, it is possible that red algae other than *Porphyra* possess active *GalEa* elements. The identification of a *PyRE10G*-related element in other red algae will provide valuable information about the evolution of GalEa elements.

Figure 7 shows a schematic representation of the phylogenetic distribution of *copia* and *gypsy* retrotransposons, LINEs, and SINEs in algae. The genome projects of all of these species other than *Porphyra* have been completed. Of the species, only *C. reinhardtii* has all of these retrotransposons. Because the number of repetitive elements, including transposable elements, is positively correlated with genome size, it is likely that the unicellular algae with smaller genomes (i.e., *T. pseudonana, Ostreococcus tauri,* and *C. merolae*) have a simple composition in terms of retrotransposon species. The absence of SINE elements in the brown alga *E. siliculosus* is an unexpected feature, because it has a larger genome (214 Mb) than *Chlamydomonas* and has LINE elements with the native structure. Because SINEs are strictly dependent on autonomous elements, such as LINEs, the combination of the absence of SINEs and the presence of LINEs observed in *E. siliculosus* can be explained as the failure of a SINE element to develop in *E. siliculosus* yet. On the other hand, the reverse combination, as observed in *Porphyra*, seems improbable. Whole genome analysis of *Porphyra* will answer this question.

Because the red alga *C. merolae* and the green alga *O. tauri* have the same genome size and have a simple composition of retrotransposons, they are simply grouped as the counterparts of retrotransposon-abundant species. However, they are distinct from each other from the perspective of the evolution of retrotransposons. Several models have been used to explain the evolution of retrotransposons.

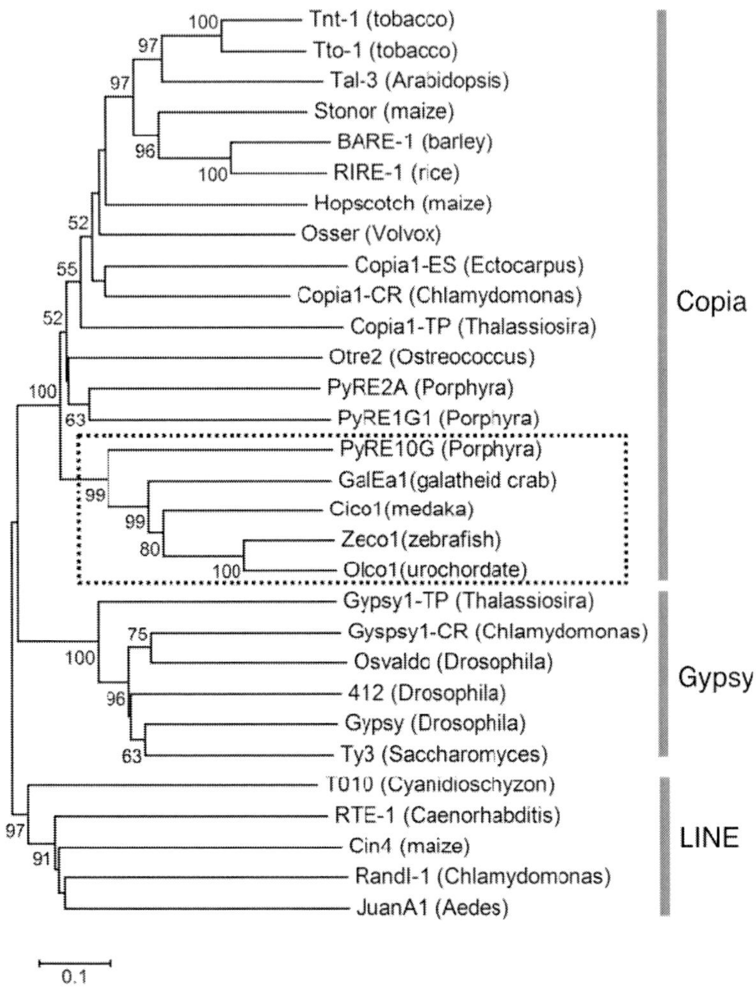

Figure 6. Phylogenetic tree of the RT domains of *Porphyra copia*-like elements and other sources. Neighbor-joining phylogram analysis based on the amino acid sequences of RT. Bootstrap values above 50% are indicated at the nodes. Accession numbers of the genes are as follows: PyRE1G1 (AB371726), PyRE10G (AB286055), PyRE2A (AB248913), Osser (X69552), Tnt1 (X13777), Stonor (AF082130), Ty3 (M23367), Gypsy (M12927), Osvaldo (AJ133521), Tto1 (D83003), Tal-3 (X13291), BARE-1 (Z17327), RIRE-1 (D85597), Hopscotch (U12626), 412 (X04132), GalEa1 (EU097705), Cico1 (DQ913003), Olco1 (DQ913000), Zeco1 (DQ913001), RTE-1 (AF054983), JuanA1 (M95171), Cin4 (Y00086). T010 is the RT gene in chromosome 20 in *Cyanidioschyzon merolae* (Matsuzaki et al., 2004; Nozaki et al., 2007). The sequences of *Chlamydomonas* elements (Copia1-CR and Gypsy1-CR) and *Thalassiosira* elements (Copia1-TP and Gypsy1-TP) were deposited in Repbase (Kohany et al., 2006). Otre2 is the putative *copia* retrotransposon with complete gene structure in chromosome 8 of

Ostreococcus tauri (Derelle et al., 2006). Coia1-ES is the putative *copia* retrotransposon with the 5520 bp complete gene structure in the supercontig-42 of *Ectocarpus siliculosus* (Cock et al., 2010).

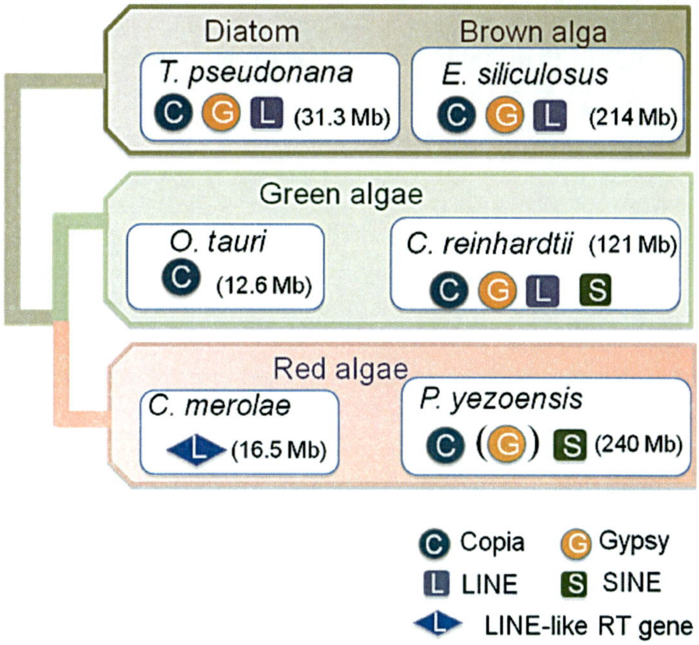

Figure 7. Phylogenetic distribution of LTR and non-LTR retrotransposons among algae. The capital letters C and G indicate the complete element of *copia* and *gypsy* retrotransposon, respectively. L and S indicate LINE and SINE element, respectively. In *C. merolae*, L means the LINE-like RT gene. The figure in parentheses shows the genome size.

Non-LTR retrotransposons are believed to be more primitive than LTR retrotransposons in that the sequences of RT (Xiong and Eickbush, 1990) and RNase H (Malik and Eickbush, 2001) of non-LTR retrotransposons are more closely related to those of bacterial elements. From the perspective of the evolution of retrotransposons, the composition of the retrotransposon in *C. merolae* is more primitive than that of *O. tauri.* Thus, a detailed description of the transposable elements and repetitive sequences in *C. merolae* is important for a discussion of the evolution of LINEs in red algae.

The *C. merolae* genome contains 24 RT-related elements (Nozaki et al., 2007). Of these, 23 elements are closely related to each other at the nucleotide sequence level, suggesting that they are members of the same subfamily. The transcripts of three elements contained an intact RT ORF. The other elements seem to be defective or dead because there is no ORF in the transcripts and repeated translation stop codons in the genome. The one distinct element is transcriptionally active, but also seems to be defective, because it has a truncated ORF in the transcript. Because all of the RT elements are unrelated to the RT of LTR retrotransposons, it can be concluded that *C. merolae* lacks LTR retrotransposons. The RT sequences of the three intact ORFs are most closely related to that of the RTE retrotransposons of LINEs. LINEs usually encode endonuclease genes upstream or downstream from the RT gene in the same frame or with a frame shift. However, the *C. merolae* RT genes lack the endonuclease gene.

Thus, the three retrotransposon-like genes are single RT genes, which are related to LINE. This is supported by the phylogenetic tree constructed from the RT sequences (Figure 6).

In addition to retrotransposons, transposons are another repetitive element found in the eukaryotic genome. The *C. merolae* genome contains only nine transposase-related elements. All nine are weakly related to the transposase sequences of the bacterial transposon insertion sequence. Unlike RT sequences, all of the transposase sequences seem to be defective, because of the presence of a stop codon within the ORF in their transcripts or genomic sequences.

The total transposable elements in the *C. merolae* genome are estimated to constitute only 0.1% of the entire genome, assuming that the average gene size of RT- and transposase-related elements are 4 and 3 kbp, respectively. In contrast to the low amount of transposable elements, 328 copies of a repetitive element containing sequences related to a shrimp white spot syndrome virus (WSV) gene were present in the *C. merolae* genome. A BLASTX search suggested that all of these elements were weakly related to a putative protein encoded in the shrimp WSV genome, the WSV486 protein. However, all of the sequences contained a truncated ORF. Although the genome of WSV has 181 ORFs (Yang et al., 2001), no sequence related to the WSV gene other than the WSV468 gene was present in the *C. merolae* genome. Thus, the origin of these elements remains unclear. These repetitive elements have an average size of 2.6 kbp and are distributed randomly on all chromosomes. The total amount of these elements is estimated to constitute about 5% of the genome. This is comparable to the amount of transposable elements in yeasts, which have the smallest amount of transposable elements (Kazazian, 2003). Thus, these repetitive elements are the major component of the repetitive DNA in the *C. merolae* genome. Most of the repetitive elements were not expressed, although some elements were cotranscribed with a flanking gene. Interestingly, an approximately 600-bp sequence at the 3'-end of the three RT transcripts with the complete RT ORF was very similar to a repetitive element, and 16 defective RT elements also contained the sequences of the repetitive elements. This suggests that these RT elements were involved in the amplification of the repetitive element. If these RT genes are primitive elements of LINEs, the endonuclease genes might exist to help the propagation of these RT genes, because these RT genes lack endonuclease genes. Based on the sequence similarity of the integrases of LTR retrotransposons, the endonucleases of non-LTR retrotransposons, and DNA transposases, Capy et al. (1996) postulated that the integrase of LTR retrotransposons originated from bacterial insertional elements. According to the model by Malik and Eickbush (1999, 2001) and Malik (2005), in the early stage of the evolution of non-LTR retrotransposons, the transposase gene was fused with the RT/RNase H gene; next, the capture of the transposase upstream of the RT/RNase H gene evolved to give the extant *copia* retrotransposons, while an insertion downstream of RT/RNase H gave rise to the *gypsy* retrotransposons. In these evolutionary models, the RT/RNase H gene was described as a putative element, and no corresponding gene is present among extant elements. However, the three RT genes in *C. merolae* may be these missing elements. If a single RT gene can propagate, transposases might help to integrate these genes. Based on the gene structure, all of the transposase genes in *C. merolae* are defective. It would be interesting to test whether the introduction of an active transposase gene into *C. merolae* induced the transposition of RT genes.

We did not find any sign of LTR retrotransposons in the *C. merolae* genome. In contrast, *P. yezoensis* has *copia* retrotransposons. From the genomic Southern analysis, *gypsy* retrotransposons are abundant in the *P. yezoensis* genome (Figure 2A, B), although the full-length gene has not been reported. To discuss the evolution of retrotransposons in red algae, secondary symbionts such as diatoms and brown algae are also important because they are believed to have acquired their plastids via secondary endosymbiogenesis, in which a red algal endosymbiont was engulfed and domesticated by a non-photosynthetic eukaryotic host. Secondary endosymbiosis has given rise to diverse groups of algae, including brown algae and diatoms. Now, the genome sequences of the diatoms *T. pseudonana* (Armbrust et al., 2004) and *Phaeodactylum tricornutum* (Bowler et al., 2008), and the brown alga *E. siliculosus* (Cock et al., 2010) are available. However, no sequence homologous to *Porphyra copia*-like elements (*PyRE2A, PyRE1G1*, and *PyRE10G*) is present in these secondary symbionts. Thus, currently the relationship between *Porphyra* elements and *copia* retrotransposons in secondary symbionts remains unclear. However, all of the *copia* retrotransposons in green algae and the secondary symbionts group with *PyRE1G1* and *PyRE2A* and are separated from the GalEa group, including *PyRE10G*. The same results have been obtained using other *copia* retrotransposons in *C. reinhardtii*, *T. pseudonana*, and *E. siliculosus*. Thus, whole-genome analyses of *Porphyra* will provide information about the evolution of retrotransposons through secondary symbiosis.

7. Conclusion

Presently, the available information about retrotransposons in *P. yezoensis* has been derived from a small number of elements. However, each element seems to be a key component for discussing the evolution of plant retrotransposons. In addition to retrotransposons, transposons are a major transposable element in eukaryotes and have provided valuable tools for basic and applied studies in plant biology. However, this chapter focuses on retrotransposons from *P. yezoensis* because no full-length transposon has been identified from *P. yezoensis*. Based on an EST database, it is predicted that *P. yezoensis* possesses two types of transposon: bacterial insertional elements and eukaryotic transposons (Kazusa DNA research Institute). This is distinct from *C. merolae*, in which most transposases are related to bacterial insertional elements. Like the situation of retrotransposons, the identification of active transposons will provide important alternate tools for studying *Porphyra*. Further, transposons are predicted to be involved in the evolution of retrotransposons, especially at an early stage, as described above. Thus, the characterization of transposons will provide valuable information about the evolution of retrotransposons in *Porphyra*.

Acknowledgments

I would like to thank previous and present coworkers for their support. I also thank Dr. Naotsune Saga of Hokkaido University, Japan, for providing the gametophyte of *P. yezoensis*.

REFERENCES

Armbrust, E. V., Berges, J. A., Bowler, C., Green, B. R., Martinez, D., Putnam, N. H., Zhou, S., Allen, A. E., Apt, K. E., Bechner, M., Brzezinski, M. A., Chaal, B. K., Chiovitti, A., Davis, A. K., Demarest, M. S., Detter, J. C., Glavina, T., Goodstein, D., Hadi, M. Z., Hellsten, U., Hildebrand, M., Jenkins, B. D., Jurka, J., Kapitonov, V. V., Kroger, N., Lau, W. W., Lane, T. W., Larimer, F. W., Lippmeier, J. C., Lucas, S., Medina, M., Montsant, A., Obornik, M., Parker, M. S., Palenik, B., Pazour, G. J., Richardson, P. M., Rynearson, T. A., Saito, M. A., Schwartz, D. C., Thamatrakoln, K., Valentin, K., Vardi, A., Wilkerson, F. P. and Rokhsar, D. S. (2004) The genome of the diatom *Thalassiosira pseudonana*: ecology, evolution, and metabolism. *Science,* 306, 79–86.

Asamizu, E., Nakajima, M., Kitade, Y., Saga, N., Nakamura, Y. and Tabata, S. (2003) Comparison of RNA expression profiles between the two generations of *Porphyra yezoensis* (Rhodophyta), based on expressed sequence TAG frequency analysis. *J. Phycol.*, 39, 923-930.

Bowler, C., Allen, A., Badger, J., Grimwood, J., Jabbari, K., Kuo, Y.-C., Maheswari, U., Martens, C., Maumus, F., Otillar, R., Rayko, E., Salamov, A., Vandepoele, K., Beszteri, B., Gruber, A., Heijde, M., Katinka, M., Mock, T., Valentin, K., Vérret, F., Berges, J., Brownlee, C., Cadoret, J., Chiovitti, A., Jae Choi, C., Coesel, S., Martino, A., Detter, J., Durkin, C., Falciatore, A., Fournet, J., Haruta, M., Huysman, M., Jenkins, B., Jiroutova, K., Jorgensen, R., Joubert, Y., Kaplan, A., Kröger, N., Kroth, P., Roche, J., Lindquist, E., Lommer, M., Martin-Jézéquel, V., Lopez, P., Lucas, S., Mangogna, M., McGinnis, K., Medlin, L., Montsant, A., Oudot-Secq, M., Napoli, C., Obornik, M., Petit, J., Porcel, B., Poulsen, N., Robison, M., Rychlewski, L., Rynearson, T., Schmutz, J., Schnitzler Parker, M., Shapiro, H.J., Siaut, M., Stanley, M., J.Sussman, M., Taylor, J.S., Vardi, A., von Dassow, P., Vyverman, W., Willis, A., Wyrwicz, L., Rokhsar, D., Weissenbach, J., Armbrust, V., Green, P.J., Van de Peer, Y. and Grigoriev, I. (2008) The *Phaeodactylum* genome reveals the dynamic nature and multi-lineage evolutionary history of diatom genomes. *Nature,* 456, 239-244.

Capy. P., Vitalis, R., Langin, T., Higuet, D. and Bazin, C. (1996) Relationships between transposable elements based upon the integrase-transposase domains: Is there a common ancestor? *J. Mol. Evol.*, 42, 359-368.

Cock, J. M., Sterck, L., Rouzé, P., Scornet, D., Allen, A. E., Amoutzias, G., Anthouard, V., Artiguenave, F., Aury, J.-M., Badger, J. H., Beszteri, B., Billiau, K., Bonnet, E., Bothwell, J. H., Bowler, C., Boyen, C., Brownlee, C., Carrano, C. J., Charrier, B., Cho, G. Y., Coelho, S. M., Collén, J., Corre, E., Silva, C. D., Delage, L., Delaroque, N., Dittami, S. M., Doulbeau, S., Elias, M., Farnham, G., Gachon, C. M. M., Gschloessl, B., Heesch, S., Jabbari, K., Jubin, C., Kawai, H., Kimura, K., Kloareg, B., Küpper, F. C., Lang, D., Bail, A. L., Leblanc, C., Lerouge, P., Lohr, M., Lopez, P. J., Martens, C., Maumus, F., Michel, G., Miranda-Saavedra, D., Morales, J., Moreau, H., Motomura, T., Nagasato, C., Napoli, C. A., Nelson, D. R., Nyvall-Collén, P., Peters, A. F., Pommier, C., Potin, P., Poulain, J., Quesneville, H., Read, B., Rensing, S. A., Ritter, A., Rousvoal, S., Samanta, M., Samson, G., Schroeder, D. C., Ségurens, B., Strittmatter, M., Tonon, T., Tregear, J. W., Valentin, K., von Dassow, P., Yamagishi, T., Van de Peer, Y. and

Wincker, P. (2010) The *Ectocarpus* genome and the independent evolution of multicellularity in brown algae. *Nature*, 465, 617-621.

Deragon, J. M., Landry, B. S., Pélissier, T., Tutois, S., Tourmente, S. and Picard, G. (1994) An analysis of retroposition in plants based on a family of SINEs from *Brassica napus*. *J. Mol. Evol.*, 39, 378-386.

Derelle, E., Ferraz, C., Rombauts, S., Rouzé, P., Worden, A. Z., Robbens, S., Partensky, F., Degroeve, S., Echeynié, S., Cooke, R., Saeys, Y., Wuyts, J., Jabbari, K., Bowler, C., Panaud, O., Piégu, B., Ball, S. G., Ral, J.-P., Bouget, F.-Y., Piganeau, G., Baets, B. D., Picard, A., Delseny, M., Demaille, J., Van de Peer, Y. and Moreaua, H. (2006) Genome analysis of the smallest free-living eukaryote *Ostreococcus tauri* unveils many unique features. *Proc. Natl. Acad. Sci. USA*, 103,11647-11652.

Fawcett, J. A., Kawahara, T., Watanabe, H. and Yasui, Y. (2006) A SINE family widely distributed in the plant kingdom and its evolutionary history. *Plant Mol. Biol.*, 61, 505-514.

Finnegan. E. J, Genger R. K., Peacock, W. J. and Dennis, E. S. (1998) DNA methylation in plants. *Annu. Rev. Plant Physiol. Plant Mol. Biol.*, 49, 223-247.

Fukuda, T., Sakai, M., Takano, H., Ono, K. and Takio, S. (2004) Hypermethylation of retrotransposons in the liverwort *Marchantia paleacea* var. *diptera*. *Plant Cell Rep.*, 22, 594-598.

Galli, G., Hofstetter, H. and Birnstiel, M. L. (1981) Two conserved sequence blocks within eukaryotic tRNA genes are major promoter elements. *Nature*, 294, 626-631.

Grandbastien, M.-A. (1998) Activation of plant retrotransposons under stress conditions. *Trends Plant Sci.*, 3, 181-187.

Grandbastien, M.-A., Lucas, H., Morel, J.-B., Mhiri, C., Vernhettes, S. and Casacuberta, J. M. (1997) The expression of the tobacco *Tnt1* retrotransposon is linked to plant defense responses. *Genetica*, 100, 241-252.

Grandbastien, M.-A. Spielmann, A. and Caboche, C. (1989) Tnt1, a mobile retroviral-like transposable element of tobacco isolated by plant cell genetics. *Nature*, 337, 376-380.

Hattman, S., Kenny, C., Berger, L. and Pratt, K. (1978) Comparative study of DNA methylation in three unicellular eucaryotes. *J. Bacteriol.*, 135, 1156-1157

Hirochika, H. (1993) Activation of tobacco retrotransposons during tissue culture. *EMBO J.*, 12, 2521-2528.

Hirochika, H. (2001) Contribution of the *Tos17* retrotransposon to rice functional genomics. *Curr. Opin. Plant Biol.*, 4, 118-122.

Hirochika, H., Sugimoto, K., Otsuki, Y., Tsugawa, H. and Kanda, M. (1996) Retro-transposons of rice involved in mutations induced by tissue culture. *Proc. Natl. Acad. Sci. USA*, 93, 7783-7788.

Kapraun, D. F., Hinson, T. K. and Lemus, A. J. (1991) Karyology and cytophotometric estimation of inter-and intraspecific nuclear DNA variation in four species of *Porphyra* (Rhodophyta). *Phycologia*, 30, 458-466.

Kazazian, H. H. Jr. (2003) Mobile elements: Drivers of genome evolution. *Science*, 303,1226-1632.

Khan, H., Kozera, C., Curtis, B. A., Bussey, J. T., Theophilou, S., Bowman, S. and Archibald, J. M. (2007) Retrotransposons and tandem repeat sequences in the nuclear genomes of cryptomonad algae. *J. Mol. Evol.*, 64, 223-236.

Kohany, O., Gentles, A. J., Hankus, L. and Jurka, J. (2006) Annotation, submission and screening of repetitive elements in Repbase: RepbaseSubmitter and Censor. *BMC Bioinformatics*, 7, 474.

Kumar, A. and Bennetzen, J. L. (1999) Plant retrotransposons. *Annu. Rev. Genet.*, 33, 479-532.

Kumar, A. and Hirochika, H. (2001) Applications of retrotransposons as genetic tools in plant biology. *Trends Plant Sci.*, 6, 127-134.

Le Gall, Y., Brown, S., Marie, D., Mejjad, M. and Kloareg, B. (1993) Quantification of nuclear DNA and G-C content in marine macroalgae by flow cytometry of isolated nuclei. *Protoplasma*, 173, 123-132.

Lenoir, A., Lavie, L., Prieto, J.-L., Goubely, C., Cote J.-C., Pélissier, T. and Deragon, J.-M. (2001) The evolutionary origin and genomic organization of SINEs in *Arabidopsis thaliana*. *Mol. Biol. Evol.*, 18, 2315-2322.

Maichele, A. J., Farwell, N. J. and Chamberlain, J. S. (1993) A B2 repeat insertion generates alternate structures of the mouse muscle γ-phosphorylase kinase gene. *Genomics*, 16, 139-149.

Malik, H. S. (2005) Ribonuclease H evolution in retrotransposable elements. *Cytogenet. Genome Res.*, 110, 392-401.

Malik, H. S. and Eickbush, T. H. (1999) Modular evolution of the integrase domain in the Ty3/Gypsy class of LTR retrotransposons. *J. Virol.*, 73, 5186-5190.

Malik, H. S. and Eickbush, T. H. (2001) Phylogenetic analysis of ribonuclease H domains suggests a late, chimeric origin of LTR retrotransposable elements and retroviruses. *Genome Res.*, 11, 1187-1197.

Matsuzaki, M., Misumi, O., Shin-I, T., Maruyama, S., Takahara, M., Miyagishima, S.Y., Mori, T., Nishida, K., Yagisawa, F., Nishida, K., Yoshida, Y., Nishimura, Y., Nakao, S., Kobayashi, T., Momoyama, Y., Higashiyama, T., Minoda, A., Sano, M., Nomoto, H., Oishi, K., Hayashi, H., Ohta, F., Nishizaka, S., Haga, S., Miura, S., Morishita, T., Kabeya, Y., Terasawa, K., Suzuki, Y., Ishii, Y., Asakawa, S., Takano, H., Ohta, N., Kuroiwa, H., Tanaka, K., Shimizu, N., Sugano, S., Sato, N., Nozaki, H., Ogasawara, N., Kohara, Y. and Kuroiwa, T. (2004) Genome sequence of the ultrasmall unicellular red alga *Cyanidioschyzon merolae* 10D. *Nature*, 428, 653-657.

Mazier, M., Botton, E., Flamain, F., Bouchet J-P., Courtial, B., Chupeau, M-C., Chupeau, Y., Maisonneuve, B. and Lucas, H. (2007) Successful gene tagging in lettuce using the *Tnt1* retrotransposon from tobacco. *Plant Physiol.*, 144, 18-31.

Merchant, S. S., Prochnik, S. E., Vallon, O., Harris, E. H., Karpowicz, S. J., Witman, G. B., Terry, A., Salamov, A., Fritz-Laylin, L. K., Maréchal-Drouard, L., Marshall, W. F., Qu, L.-H., Nelson, D. R., Sanderfoot, A. A., Spalding, M. H., Kapitonov, V. V., Ren, Q., Ferris, P., Lindquist, E., Shapiro, H., Lucas, S. M., Grimwood, J., Schmutz, J., *Chlamydomonas* Annotation Team, JGI Annotation team, Grigoriev, I. V., Rokhsar, D. S. and Grossman, A. R. (2007) The *Chlamydomonas* genome reveals the evolution of key animal and plant functions. *Science*, 318, 245 -250.

Michel, D., Chatelain, G., Mauduit, C., Benahmed, M. and Brun, G. (1997) Recent evolutionary acquisition of alternative pre-mRNA splicing and 3' processing regulations induced by intronic B2 SINE insertion. *Nucleic Acids Res.*, 25, 3228-3234.

Miyao, A., Iwasaki, Y., Kitano, H. , Itoh, J., Maekawa, M. , Murata, K., Yatou, O. , Nagato, Y. and Hirochika, H. (2007) A large-scale collection of phenotypic data describing an insertional mutant population to facilitate functional analysis of rice genes. *Plant Mol. Biol.*, 63, 625-635.

Mount, S. M. and Rubin, G. M. (1985) Complete nucleotide sequence of the *Drosophila* transposable element copia: Homology between copia and retroviral proteins. *Mol. Cell Biol.*, 5, 1630-1638.

Myouga, F., Tsuchimoto, S., Noma, K., Ohtsubo, H. and Ohtsubo, E. (2001) Identification and structural analysis of SINE elements in the *Arabidopsis thaliana* genome. *Genes Genet. Syst.*, 76, 169-179.

Nikaido, I., Asamizu, E., Nakajima, M., Nakamura, Y., Saga, N. and Tabata, S. (2000) Generation of 10,154 expressed sequence tags from a leafy gametophyte of a marine red alga, *Porphyra yezoensis*. *DNA Res.*, 7, 223-227.

Noutoshi, Y., Ito, Y., Kanetani, S., Fujie, M., Usami, S. and Yamada, T. (1998) Molecular anatomy of a small chromosome in the green alga *Chlorella vulgaris*. *Nucleic Acids Res.*, 26, 3900-3907.

Nozaki, H., Takano, H., Misumi, O., Terasawa, K., Matsuzaki, M., Maruyama, S., Nishida, K., Yagisawa, F., Yoshida, Y., Fujiwara, T., Takio, S., Tamura, K., Chung, S.J., Nakamura, S., Kuroiwa, H., Tanaka, K., Sato, N. and Kuroiwa, T. (2007) A 100%-complete sequence reveals unusually simple genomic features in the hot-spring red alga *Cyanidioschyzon merolae*. *BMC Biol.*, 5, 28.

Okada, N., Hamada, M., Ogiwara, I. and Ohshima, K. (1997) SINEs and LINEs share common 39 sequences: a review. *Gene*, 205, 229-243.

Peddigari, S., Zhang, W., Sakai, M., Takechi, K., Takano, H. and Takio, S. (2008a). A *copia*-like retrotransposon gene encoding *gypsy*-like integrase in a red alga, *Porphyra yezoensis*. *J. Mol. Evol.*, 66,72-79.

Peddigari, S., Zhang, W., Takechi, K., Takano, H. and Takio, S. (2008b) Two different clades of *copia*-like retrotransposons in the red alga, *Porphyra yezoensis*. *Gene*, 424,153-158.

Peterson-Burch, B. D. and Voytas, D. F. (2002) Genes of the pseudoviridae (Ty1/*copia* retrotransposons). *Mol. Biol. Evol.*, 19, 1832-1845.

Pouteau, S., Huttner, E., Grandbastien, M. A. and Caboche, M. (1991) Specific expression of the tobacco Tnt1 retrotransposon in protoplasts. *EMBO J.*, 10, 1911-1918.

Pouch-Pélissier, M. N., Pélissier, T., Elmayan, T., Vaucheret, H., Boko, D., Jantsch, M.F. and Deragon, J.M. (2008). SINE RNA induces severe developmental defects in *Arabidopsis thaliana* and interacts with HYL1 (DRB1), a key member of the DCL1 complex. PLoS Genet 4, e1000096.10.1371/journal.pgen.1000096.

Ramallo, E., Kalendar, R. , Schulman, A. H. and Martínez-Izquierdo, J. A. (2008) *Reme1*, a *copia* retrotransposon in melon, is transcriptionally induced by UV light. *Plant Mol. Biol.*, 66, 137-150.

Schmidt, T. (1999) LINEs, SINEs and repetitive DNA: non-LTR retrotransposons in plant genomes. *Plant Mol. Biol.*, 40, 903–910.

Takeda, S., Sugimoto, K., Otsuki, H. and Hirochika, H. (1998) Transcriptional activation of the tobacco retrotransposon *Tto1* by wounding and methyl jasmonate. *Plant Mol. Biol.*, 36, 365-376.

Terrat, Y., Bonnivard, E. and Higuet, D. (2007) GalEa retrotransposons from galatheid squat lobsters (decapoda, Anomura) define a new clade of *Ty1/copia*-like elements restricted to aquatic species. *Mol. Genet. Genomics*, 279, 63-73.

Umeda, M., Ohtsubo, H. and Ohtsubo, E. (1991) Diversification of the rice *Waxy* gene by insertion of mobile DNA elements into introns. *Jpn. J. Genet.*, 66, 569-586.

Voinett, O. (2009) Origin, biogenesis, and activity of plant microRNAs. *Cell*, 136, 669-687.

Wang, D., Lin, X., Hirayama, K., Li, Z., Ohno, T., Zhang, W., Namihira, T., Katsuki, S., Takano, H., Takio, S. and Akiyama, H. (2010) A new application of underwater pulsed streamer-like discharge to transcriptional activation of retrotransposon of *Porphyra yezoensis*. *IEEE Trans Plasma Sci.*, 38, 39-46.

Xiong, Y. and Eickbush, T. H . (1990) Origin and evolution of retroelements based upon their reverse transcriptase sequences. *EMBO J.*, 9, 3353-3356.

Xu, J. H., Osawa, I., Tsuchimoto, S., Ohtsubo, E. and Ohtsubo, H. (2005) Two new SINE elements, *p-SINE2* and *p-SINE3*, from rice. *Genes Genet. Syst.*, 80, 161-171.

Yang, F., He, J., Lin, X. H., Li, Q., Pan, D., Zhang, X. B. and Xu, X. (2001) Complete genome sequence of the shrimp white spot bacilliform virus. *J. Virol.*, 75, 11811-11820.

Yoshioka, Y., Matsumoto, S., Kojima, S., Ohshima, K., Okada, N. and Machida, Y. (1993) Molecular characterization of a short interspersed repetitive element from tobacco that exhibits sequence homology to specific tRNAs. *Proc. Natl. Acad. Sci. USA*, 90, 6562-6566.

Zhang, W., Sakai, M., Lin, X., Takechi, K., Takano, H. and Takio, S. (2006) Reverse transcriptase-like sequences related to retrotransposon in a red alga, *Porphyra yezoensis*. *Biosci. Biotech. Biochem.*, 70, 1999-2003

Zhang, W., Lin, X., Peddigari, S., Tackechi, K., Takano, H. and Takio, S. (2007) Characterization of short interspersed elements (SINEs) in a red alga, *Porphyra yezoensis*. *Biosci. Biotech. Biochem.*, 71, 618-622.

In: Porphyra yezoensis
Editor: Koji Mikami

ISBN 978-1-61122-304-0

Chapter 5

COLLECTION, ANALYSIS, AND UTILIZATION OF *PORPHYRA YEZOENSIS* EXPRESSED SEQUENCE TAGS

***Erika Asamizu*[1,4]*, Yukihiro Kitade*[2,3] *and Satoshi Tabata*[4*]**

[1] Gene Research Center, University of Tsukuba, 1-1-1 Tennodai, Tsukuba, Ibaraki 305-8572, Japan

[2] Faculty of Fisheries Sciences, Hokkaido University, Hakodate, Hokkaido 041-8611, Japan

[3] Research Institute of Innovative Technology for the Earth (RITE), 9-2 Kizugawadai, Kizugawa, Kyoto 619-0292, Japan

[4] Kazusa DNA Research Institute, 2-6-7 Kazusa-kamatari, Kisarazu, Chiba 292-0812, Japan

ABSTRACT

Expressed sequence tags (ESTs) are partial cDNA sequences that allow compilation of the gene constituents in a genome in a cost-effective manner, and a large number of ESTs have been intensively collected for a wide variety of organisms. A total of 22,069 ESTs have been accumulated from both the gametophyte and sporophyte phases of *Porphyra yezoensis* and made available at the public DNA databases. These ESTs have been utilized for identification and isolation of candidate genes, which are involved in asexual sporulation, demonstrating the usefulness of the EST information and the corresponding cDNA clones to understand the genetic systems of this alga.

INTRODUCTION

Porphyra yezoensis is an attractive model organism for basic and applied studies of marine plants due to several key features. Its life cycle is quite short (only a few months under

* Author for correspondence: e-mail tabata@kazusa.or.jp

laboratory conditions) (Kuwano et al., 1996), and its genome is estimated to be relatively small (2.6×10^8 bp) (Kapraun et al., 1991; Le Gall et al., 1993) and comparable in size to those of Arabidopsis thaliana and Oryza sativa. P. yezoensis has two contrasting heteromorphic life cycle stages, a macroscopic haploid foliate gametophyte phase and a microscopic diploid filamentous sporophyte phase. These gametophytic and sporophytic phases exhibit a number of different structural features (Cole and Conway, 1975): the chloroplast number (single or multiple), growth mechanism (diffuse or apical), cell wall composition (xylan or cellulose), and plugged pit connection structure (absent or present)—all vary according to the life cycle stage. However, the molecular mechanisms underlying these features are unknown because conventional genetics approaches cannot be used for this organism.

Recent advances made in genomics technologies are expected to facilitate the development of genetic systems for organisms that to date have not been possible. For example, advances have been made in genome sequencing, gene prediction, and cDNA sequencing, as well as transcriptome, proteome, and metabolome analyses. Expressed sequence tags (ESTs), one type of cDNA sequencing, are collected in order to compile the gene constituents in a genome in a cost effective way (http://www.ncbi.nlm.nih.gov/dbEST/dbEST_summary.html). A large number of ESTs have been intensively accumulated for a wide variety of organisms, including model plants such as O. sativa, A. thaliana, liverwort, and Chlamydomonas reinhardtii. To date, a total of 22,069 P. yezoensis ESTs have been deposited in the EST database. In this chapter, ESTs collected from P. yezoensis in the gametophyte and sporophyte phases are briefly summarized. In addition, examples of how these accumulated ESTs can be utilized are described.

1. EST Analysis In Porphyra Yezoensis

Comprehensive EST analyses of *P. yezoensis* were performed to further our understanding of the genetic system in red algae (Nikaido et al., 2000; Asamizu et al., 2003). In the initial phase of this project, 10,154 ESTs were identified from a normalized, size-selected *P. yezoensis* Ueda (strain TU-1) cDNA library (Kuwano et al., 1996) generated from the gametophytic phase. To compare gene expression between the two phases of the *P. yezoensis* life cycle and correlate these differences to the observed morphological and physiological differences, 10,625 EST sequences were also generated from the sporophyte cDNA library. This chapter describes both the methods used to collect the ESTs and their features as revealed by computer analyses.

1.1. Methods for EST Collection and Analyses

The process of cDNA library construction is summarized as follows. Leafy gametophytes of the P. yezoensis Ueda strain (TU-1) were grown in a ESL medium containing 3.5% Sealife powder (Marinetech Co., Ltd., Japan) and 1% (v/v) ESS2 stock solution (pH 8.0) (Kitade et al., 1996) in a ten-hour photoperiod of fluorescent illumination (50 μmol • m-2 • s-1) at 15°C with constant air bubbling for four weeks. Filamentous sporophytes of the same P. yezoensis strain were grown in a 24-hour photoperiod of fluorescent illumination (17 μmol • m-2 • s-1) at 25°C in the same medium described above with constant air bubbling for four to eight weeks. Total RNA was extracted using a modified SDS/phenol method (Nikaido et al., 2000). Two types of cDNA libraries were made to obtain long clones in an intact form: normalized and size-selected. Normalization was performed for the library containing 1–3 kb fragments as described by Bonaldo et al. (1996). This method is based on self-hybridization between a single-stranded library generated by the combined action of the phage F1 endonucleases II and III and an excess amount of cDNA inserts generated by polymerase chain reaction (PCR). The remaining single-stranded circles were purified using hydroxyapatite chromatography, converted into double-strand DNA and transformed into Escherichia coli. The cDNA library that contained fragments over 3 kb was called the size-selected library. This library was used without normalization to avoid physical degradation of the cDNA fragments during the normalization process.

The 5' ends of the cDNA clones were sequenced to generate the ESTs using the Sanger sequencing protocol and a fluorescent DNA sequencer (models 377 and 3730, Applied Biosystems, USA). The vector-derived sequence and ambiguous sequences were removed from the collected EST sequences prior to the computer-aided analyses. To identify the number of independent EST species, clustering of the EST sequences was performed. The end sequences were compared to the entire sequence dataset using the BLASTN program (Altschul et al., 1997), and clones that showed over 95% identity for more than 100 bp were included in the same group.

Each unique sequence from non-redundant EST groups was translated into its amino acid sequence in all six reading frames and subjected to a similarity search against the non-redundant NCBI protein database (nr) using the BLASTX program. Similarity between a deduced amino acid sequence and a known sequence was judged to be significant when the E-value was less than 1.0-14. When no similarity was detected at the amino acid level, the unique sequence was subjected to a similarity search against the non-redundant nucleotide sequence database (nt) using the BLASTN program, and an optimal score of greater than 130 was judged to be significant.

1.2. EST Collection from Gametophytes and Sporophytes

Single-pass sequencing from the 5' end of the cDNA was performed for 10,154 clones from both the normalized and size-selected libraries. The average length of these ESTs was 470 bp and the average GC content was 65.2%. One notable feature of these sequences was that codons with a G/C at the third position occurred frequently: the GC content of the third position (79.4%) was much higher than that at the first (62.2%) and second (45.0%) positions. This suggests that the genome of *P. yezoensis* has been exposed to high GC pressure during evolution. In addition, these base preference characteristics may be useful for predicting protein-coding regions in the *P. yezoensis* genome.

Clustering of the EST sequences revealed a total of 3,267 non-redundant EST groups, of which 1,127 were generated by clustering 8,014 ESTs, while 2,140 were unique. It should be noted, however, that the number of EST species does not indicate the number of corresponding genes because non-redundant ESTs can originate from different regions of a single gene. At the time of publication (Nikaido et al., 2000), 1,080 (33.1%) of the 3,267 non-redundant ESTs showed sequence similarity to genes registered in public databases, and

2,187 (66.9%) were novel. Similar results were reported in an EST analysis of the green algae *C. reinhardtii*, where 2,616 (76.2%) out of 3,433 non-redundant EST groups were novel (Asamizu et al., 1999). By contrast, only 2,692 (37.7%) out of 7,137 non-redundant EST groups were novel in an analysis of the legume *Lotus japonicus* (Asamizu et al., 2000). The high percentage of new sequences identified for *P. yezoensis* and *C. reinhardtii* may be due to the limited number of the sequences that have been registered in the public DNA databases for red and green algae relative to higher plants. Alternatively, since the Rhodophyta group, which includes *P. yezoensis*, and the Viridiplantae group (green plants) are evolutionarily distant, group specific genes and/or genes with highly divergent sequences may be present.

A total of 10,625 sporophyte cDNA clones (3,783 clones from the normalized library and 6,842 clones from the size-selected library) were sequenced from the 5' end. These ESTs were clustered into 2,556 non-redundant groups. The quality of the libraries with respect to the coverage of the 5'-terminus was assessed by comparing the 5'-end sequences to sequences of known proteins. One hundred clones each were chosen from both the normalized library and the size-selected library, of which 50 (50%) and 54 (54%), respectively, were found to contain the translation initiation codon. These data indicate that roughly half of the cDNA clones were full-length.

1.3. Data Analyses

To identify the unique EST groups in the two growth phases, the sporophyte and the gametophyte ESTs were grouped together for a total of 20,779 EST sequences that clustered into 4,496 non-redundant groups. Among them, 1,013 (22.5%) groups were classified as ESTs that commonly occurred in both generations, whereas a relatively large number of groups appeared only in either the sporophyte or gametophyte EST population [1,543 (34.3%) and 1,940 (43.1%) groups, respectively].

The 4,496 non-redundant groups were compared with the public protein database using the BLASTX program (Altschul et al., 1997). Of these, 1,451 (32.3%) groups showed significant ($P<1.0^{-5}$) similarity to genes with known or putative functions; 301 (6.7%) showed similarity to predicted genes in the genomes of various organisms, including *A. thaliana*, *Caenorhabditis elegans*, *Homo sapiens*, *Schizosaccharomyces pombe* and *Synechocystis* sp. PCC6803, etc.; and the remaining 2,744 (61.0%) groups showed no similarity with known or predicted protein sequences. ESTs that showed sequence similarity to genes with known or putative functions were grouped into 18 categories with respect to their different biological roles generally as described by Riley (1993). In both generations, most of the ESTs fell into three major categories: energy metabolism, protein fate, and protein synthesis. These categories contained 2,684 and 2,887 clones from the gametophyte and the sporophyte phases, respectively, which comprises more than one quarter of the clones analyzed.

To preliminarily identify candidate genes that were differentially expressed in the gametophytic and the sporophytic phases, we compared the frequency with which each EST group appeared. The frequency data were evaluated using the statistical significance test described by Audic and Claverie (1997). Eighty-nine of the EST groups were more abundantly represented in the gametophyte phase than the sporophyte phase ($P< 0.01$). Among these, 37 showed similarity to genes with known or putative functions, 14 were similar to hypothetical genes, and the remaining 38 showed no similarity. In the sporophyte phase, 112 EST groups were represented more abundantly. Of these, 69 showed similarity to genes with known or putative functions, nine were similar to hypothetical genes, and 34 showed no similarity.

Table 1 List of the gene categories that exhibited significant (*P*<0.01) differences in EST abundance between the gametophyte phase and the sporophyte phase

A. Gametophyte > Sporophyte

Molecular function	G[a]	S[b]
Cellular structure, organization, and biogenesis	28	1
Central intermediary metabolism	7	0
DNA metabolism	319	28
Energy metabolism	86	25
Protein fate	91	7
Protein synthesis	699	91
Regulatory functions	38	11
Signal transduction	15	0
Transport and binding proteins	8	0

B. Sporophyte > Gametophyte

Definition of similar sequences	G[a]	S[b]
Cellular structure, organization, and biogenesis	16	173
Central intermediary metabolism	30	81
DNA metabolism	2	14
Energy metabolism	86	373
Fatty acid and phospholipid metabolism	0	13
Protein fate	106	616
Protein synthesis	116	461
Purines, pyrimidines, nucleosides, and nucleotides	0	7
Regulatory functions	0	19
Signal transduction	16	88
Transcription	4	42
Transport and binding proteins	26	216

a) Number of gametophyte ESTs
b) Number of sporophyte ESTs

A list of the putative differentially expressed genes with known or theoretical functions is shown in Table 1. Differential expression of a randomly selected subgroup of genes was confirmed by semi-quantitative reverse transcription-PCR (RT-PCR) analyses (Asamizu et al., 2003).

Since the morphological difference between the gametophyte and sporophyte phases is largely determined by the genes expressed within them, the significance value of the tag frequency in the two libraries was calculated to preliminarily identify differentially expressed gene candidates (Table 1). Of the 89 EST groups that were highly expressed in the gametophyte, 14 with similarity to ribosomal protein genes were significantly more abundant in the gametophyte than in the sporophyte phase among the 76 ribosomal protein genes identified in the analysis. In contrast, the 112 EST groups identified as highly expressed genes candidates in the sporophyte phase included various types of genes, such as tubulin genes involved in cellular structure, genes involved in diverse metabolic systems, heat shock protein genes, and ABC transporter genes. In both generations, many hypothetical and previously unknown genes were also identified. Although significantly more EST sequences were derived from the normalized libraries (11,773 of 20,779) than the size-selected libraries, semi-quantitative RT-PCR analysis suggested that this difference likely reflected distinct gene expression levels.

These analyses identified gametophyte- and sporophyte-specific ESTs, but it remains unclear whether these genes play crucial roles in determining the morphological and physiological differences between these two phases. Further functional studies employing molecular genetics approaches are needed to test the functions of these genes. In addition, it is likely that our statistical analysis failed to detect genes expressed at low levels with inadequate sequence tag numbers. Clustering of a total of 20,779 ESTs revealed that a large portion of the 4,496 non-redundant EST groups were detected only either in the sporophyte or the gametophyte EST populations (1,543 (34.3%) and 1,940 (43.1%), respectively). The majority of these phase-specific ESTs were not considered statistically significant as the sequence tag number was insufficient. In addition to the differentially represented EST groups identified in this analysis, it is thus necessary to further examine low abundance transcripts for phase-specific genes.

The EST sequences reported in this chapter appear in the GenBank/EMBL/DDBJ database (accession numbers AV429311-AV439464 and AU186563-AU197187), and the details of the search results are provided online at http://www.kazusa.or.jp/en/plant/porphyra/EST/.

2. Utilization of ESTs

2.1. Isolation of Target CDNA Clones

P. yezoensis ESTs are useful for accelerating the isolation of the corresponding cDNA clones of interest. Actin-related proteins (ARPs) are believed to be involved in various cellular functions, such as vesicle motility, mitosis, actin filament dynamics and the modulation of chromatin structure. ARPs exhibit moderate sequence identity (17–60%) to conventional actins and are grouped into several classes that are highly conserved across a wide range of eukaryotes. Unlike actin isovariants, which predominate either in vegetative or reproductive tissues, most ARPs in plants are constitutively expressed in all organs and tissues. Several cDNA sequences that exhibited similarity to the actin-related protein 4 (ARP4) gene of the terrestrial plant *A. thaliana* were found in the *P. yezoensis* EST database (http://www.kazusa.or.jp/en/plant/porphyra/EST/). The entire *P. yezoensis* ARP4 (PyARP4) open reading frame (ORF) was previously cloned and sequenced and was found to contain a putative nuclear localization signal (NLS) (Kitade et al., 2006). The class 4 ARPs have been reported to localize to the nucleus in other organisms, and thus likely contain nuclear targeting sequences. Although it has not yet been shown that the PyARP4 motif actually functions as a nuclear targeting signal *in vivo*, its presence strongly suggests that PyARP4 may be localized to the nucleus. RT-PCR analysis indicated that the levels of *PyARP4* did not differ significantly among the four *P. yezoensis* life cycle stages (two gametophyte forms and two sporophyte forms) and were lower than the levels of the conventional actin *PyACT*. Since *PyARP4* is expressed constitutively throughout the *P. yezoensis* life cycle at a low level, it may be useful in other studies as an internal standard for gene expression analyses.

Since no conventional and/or direct methods are available to elucidate gene function in this alga, we attempted to develop a gene function assay. To this end, the EST information and the corresponding cDNA clones were used to construct expression plasmids for heterologous functional complementation. This strategy uses cDNAs from one organism to restore the wild-type phenotype of another organism carrying a known mutation. Complementation analysis was carried out using an arginine-auxotroph mutant of the yeast *Saccharomyces cerevisiae* (Endo et al., 2006) deficient for the gene *ARG1*, which encodes an argininosuccinate synthetase (ASS). ASS is involved in the penultimate reaction in the arginine synthesis pathway and is found widely from bacteria to vertebrates. The *ARG1* gene of *P. yezoensis* (*PyARG1*) was obtained by PCR using a cDNA clone selected based on the EST information, and the PCR fragment was then subcloned into the yeast expression vector pYES2. Expression of *PyARG1* in an *ARG1*-deficient strain of yeast successfully complemented the mutant phenotype. The transformed cells could survive on selective medium lacking arginine, and *PyARG1* transcripts were detected by RT-PCR. This technique will provide new opportunities further investigating the functions of various genes in this alga.

EST information and corresponding cDNA clones can also be used for characterizing gene families. Indeed, ESTs served as an indispensible resource for the identification of new members of the actin gene family (Kitade et al., 2008a). Actin is a well-known cytoskeletal component with fundamental roles in both cytoplasmic and nuclear events. Bhattacharya et al. (2000) suggested that actin genes are useful markers for morphological evolution in early diverging land plants since they occur as a single copy in most green algae, as two to three copies in bryophytes, and form more complex gene families in ferns and angiosperms. In the red algae lineage, however, the correlation between evolution and the number of actin genes is unclear. For example, the Florideophycidae *Chondrus,* which has a carposporophytic, multicellular thallus and a pit plug throughout its life cycle, encodes only a single actin gene. In contrast, both genomic Southern analysis and *in silico* analysis of the *P. yezoensis* EST data revealed that Protoflorideophycidae *Porphyra*, which does not have either a carposporophytic multicellular thallus or a pit plug in the gametophytic generation, contains multiple actin genes. In fact, a total of 27 sequences showing similarity to portions of the actin gene were found in the EST database, and these sequences were classified into four groups (*PyACT1- PyACT4).* One of these groups was identical to a previously reported *P. yezoensis* actin sequence and was named *PyACT1* (accession no. AB039831). The remaining *PyACT* sequences were obtained by various methods. For *PyACT3*, the entire ORF was contained in a corresponding cDNA clone. For *PyACT2* and *4*, the entire sequences were obtained by a combination cDNA library screening and 5' rapid amplification of cDNA ends (5' RACE). The amino acid identity between the four isoforms exceeded 83%, and each contained a putative nuclear export signal (NES). The result of the sequence analysis suggested that PyACT2 functions in the nucleus as a monomer (G-actin) or in other unconventional forms. In addition, the distribution and position of the introns were different to those in the florideophycean actin genes, even though recent phylogenetic studies reported that Florideophyceae forms a monophyletic group with the Bangiales order, which includes *Porphyra*. The amount of *PyACT3* transcript was measured by semi-quantitative RT-PCR using RNA samples isolated from the four different life cycle stages (two gametophyte forms and two sporophyte forms). Additional RNA analysis was performed by Northern blotting using RNA samples harvested from the apical and basal parts of mature gametophytic thalli. The expression level of *PyACT3* was significantly higher in the mature gametophytes than that in the vegetative stage, although *PyACT3* mRNA was detected at similar levels in both parts of thalli. The expression levels of *PyACT2* and *4*, on the other hand, did not differ significantly between the mature and vegetative gametophytes. These results suggest that *PyACT3* may serve as a molecular marker for monitoring thallus maturation. Development of molecular markers for gene expression, such as *PyACT,* will contribute to quality control of "nori" harvest and accelerate *Porphyra* aquacultural science.

2.2. cDNA Macroarray

In addition to their other applications, EST clones allow genome-wide gene expression analysis using cDNA array technology. According to Schena and Davis (2000), array experimentation should begin with a biological question but not necessarily with a hypothesis. Asexual reproduction via archeospores in *P. yezoensis* gametophytes is very important in "nori" farming. However, little is known regarding the molecular basis of this developmental process. In particular, it remains unclear how marginal vegetative cells of blades differentiate into archeospores and what kinds of genes are involved in this process. To identify candidate genes involved in asexual sporulation of the gametophyte blades of *P. yezoensis*, gene expression profiles of the four different developmental life cycle stages (three from the gametophyte phase and one from the sporophyte phase), were compared using cDNA macroarrays that contained 4,896 non-redundant EST groups (Kitade et al., 2008b). In parallel, RT-PCR was performed to verify genes expressed predominantly or exclusively during asexual sporulation in gametophytes and to exclude those expressed constitutively in sporophytes. Based on the results of the array analysis, 75 clones were chosen for further investigation. Thirty-one of these clones were excluded from the candidate list since their transcripts were detected by RT-PCR in filamentous sporophytes, which do not undergo asexual sporulation. The remaining 44 clones were subjected to detailed RT-PCR analysis to confirm the stage specificity of their transcription. Differentially expressed genes were defined as those that were detected in gametophytes 5 mm in diameter that formed archeospores (5G-A), but not in filamentous sporophytes undergoing vegetative growth (S-VG). Eleven of these gametophyte (G)-specific genes were identified. Among these, two genes exhibited expression patterns highly specific to the gamytophyte phase, and these genes were designated *ASPO2608* and *ASPO1527*.

The *ASPO2608* gene contained an ORF capable of encoding a 220 amino acid protein. The ASPO2608 protein was predicted to be cysteine-rich (12.44% in the mature form) and did not exhibit sequence similarity to known or hypothetical proteins registered in public protein databases. The first 27 amino acids were predicted to constitute a signal peptide by two independent computer-aided analyses. The amino acids from C_{34} to C_{45} (CPCALSGGGGRC) were identified as an epidermal growth factor (EGF)-like domain. In addition, a putative glycosaminoglycan attachment site was found within the EGF-like domain. Although the spacing of the cysteine residues in ASPO2608 was slightly divergent from the EGF-like domain signature, the presence of the putative signal peptide and the lack of any other potential transmembrane domains suggested that this protein is likely to be localized extracellularly and potentially secreted by vegetative or nascent archeospore cells. Many growth factors function in a paracrine manner in cooperation with their receptors. Given this, ASPO2608 may be involved in a signal transduction pathway that triggers differentiation of archeospores. Alternatively, ASPO2608 may be directly integrated in the extracellular proteoglycan milieu through its glycosaminoglycan attachment site.

The *ASPO1527* gene contained an ORF encoding 394 amino acids. The first 30 amino acids were predicted to constitute a signal peptide. A putative *N*-glycosylation site was also predicted in the ASPO1527 amino acid sequence. The ASPO1527 protein did not show any sequence similarity to known or hypothetical proteins in the public protein databases. The sequence analysis suggested that the ASPO1527 protein is secreted by vegetative or nascent archeospore cells. The restricted temporal expression pattern of *ASPO1527* supports the idea that the corresponding protein is an extracellular component involved in cell adhesion to a substratum, which is an important factor in asexual sporulation.

To further analyze the 11 G-specific genes and to identify different types of asexual sporulation-related genes using the previous EST analysis, the value data (signal intensity data) from the macroarray analysis were ranked. Changes in the rank order during gametophytic generation were then investigated. For constitutively expressed housekeeping genes, such as *PyEF-1 α*, the 60S ribosomal protein L3 gene, *PyACT1*, and *PyARP4*, no remarkable change in the rank order was observed among the three gametophytic stages. Three of the G-specific genes, however, exhibited rank changes of >1,000 between the 1 mm gametophytes with no archeospores (1G-NA) and the 5G-A stage. Two of these three genes (i.e., *ASPO1527* and *ASPO2608*) did not exhibit a remarkable change in rank between the 1G-NA stage and the 20 mm gametophytes with vegetative growth (20G-VG) stage, which is similar to the housekeeping genes.

Of the 11 genes preferentially expressed in gameotophytes, six exhibited a >1,000 change in rank between the 1G-NA and 5G-A stage. Two of the six were ranked higher in the 5G-A stage, while the other four were ranked lower in the 5G-A stage. The two genes that ranked higher were categorized as "energy metabolism" and "signal transduction" genes, while the four that ranked lower fell into "energy metabolism," "protein synthesis," and "unclassified" categories. The predicted protein encoded by one of the up-regulated genes was classified as a signal transduction molecule similar to cyclase associated protein (CAP) 2 in *Xenopus laevis*. CAPs are highly conserved monomeric actin-binding proteins present in a wide range of organisms, including yeast, flies, plants, and animals. CAPs are multifunctional molecules that contain domains involved in binding actin, adenylyl cyclase (AC), and Src homology 3 (SH3) domains, and in oligomerization. However, no interaction between the AC binding domain and adenylyl cyclases has been demonstrated in higher eukaryotes. It is speculated that CAPs may sequester globular actin (G-actin) during actin disassembly or function as actin monomer delivery proteins that promote actin assembly. Although actin was not directly identified as a candidate gene in the screens described above, these findings suggest it should be a target of CAP in *P. yezoensis*.

Another of the up-regulated genes was predicted to encode an energy metabolism protein and showed the highest degree of sequence similarity to ribulose-5-phosphate-3-epimerase (RPE) in *A. thaliana*. RPE is one of several Calvin-cycle enzymes. The macroarray analysis suggested that in the 5G-A stage, the transcription level of a set of Calvin-cycle genes, including *PyRPE*, was considerably higher in the 1G-NA stage. To confirm this, northern analysis was performed for three Calvin-cycle genes, *PySBP*, *PyPGK*, and *PyFBP*. The transcriptional levels of these genes were higher in the 5G-A stage than in the 1G-NA stage, and the difference in the *PySBP* level was statistically significant. Calvin-cycle enzymes catalyze the carbon-fixation reaction in which carbon dioxide is assimilated to produce carbohydrates in chloroplast stroma. The changes in the level of Calvin-cycle gene expression may provide a clue to unravel the mechanisms involved in metabolic regulation during asexual sporulation. This regulation may involve the carbon to nitrogen (C:N) balance, as in terrestrial plants, both high carbon and low nitrogen inhibit photosynthesis. Specifically, a deficiency in nitrogen rapidly inhibits growth, induces carbohydrate accumulation, and subsequently inhibits photosynthesis. Conversely, high nitrogen and low carbon activates photosynthesis and leads to elevated Calvin-cycle gene expression. In genus *Nostoc* cyanobacterium, large quantities of extracellular polymeric substances (EPSs) consisting mainly of polysaccharides are produced only when the amount of fixed carbon exceeds the amount of nitrogen available. In addition, EPS protein-to-carbohydrate ratios in the range of 2–2.5 are characteristic for naked morphotypes, whereas this ratio is 1.5 or lower for encapsulated morphotypes (Otero and Vincenzini, 2004). It seems likely that asexual sporulation in *Porphyra* requires relatively high photosynthetic activity due to the presence of low levels of carbon and high protein-to-carbohydrate EPS ratios. In red algae, floridean starch is stored in the cytosol, whereas starch accumulates in the plastid in green algae and higher plants. At night, this starch is broken down to help support the metabolic needs of the plant. Consumption of carbon at night should result in a high protein to carbohydrate ratio. This C:N balance is consistent with the ratio that is expected in asexual sporulation in *Porphyra* because archeospores are frequently formed with a short-day photoperiod (10:14 L:D) rather than a long-day one (14:10 L:D).

CONCLUSION

This chapter focused on the characteristics of existing Porphyra ESTs and some examples of their uses. The successful identification and isolation of candidate genes, including those involved in asexual sporulation, demonstrates that the EST information and the resulting macroarray analysis can serve as useful resources to study Porphyra genetics, even though gene functions are largely uncharacterized for this alga. Further development of genomic analysis tools, including collection of additional ESTs, accumulation of transcriptome data, and generation of whole genome sequences, together with development of basic techniques, such as transformation and complementation, is expected in the near future. These technologies will aid in the establishment of *P. yezoensis* as an ideal model for the study of marine plants.

REFERENCES

Altschul, S. F., Madden T. L., Schaffer, A. A., Zhang, J., Zhang, Z., Miller, W. and Lipman, D. J. 1997. Gapped BLAST and PSI-BLAST: a new generation of protein database search programs. *Nucleic Acids Res.* 25:3389-402.

Asamizu, E., Nakamura, Y., Sato, S., Fukuzawa, H. and Tabata, S. 1999. A large scale structural analysis of cDNAs in a unicellular green alga, *Chlamydomonas reinhardtii*. I. Generation of 3433 non-redundant expressed sequence tags. *DNA Res.* 6:369-73.

Asamizu, E., Nakamura, Y., Sato, S. and Tabata, S. 2000. Generation of 7137 non-redundant expressed sequence tags from a legume, *Lotus japonicus*. *DNA Res.* 7:127-30.

Asamizu, E., Nakajima, M., Kitade, Y., Saga, N., Nakamura, Y. and Tabata, S. 2003. Comparison of RNA expression profiles between the two generations of *Porphyra yezoensis* (Rhodophyta), based on expressed sequence tag frequency analysis. *J. Phycology* 39:923-30.

Audic S. and Claverie J. M. 1997. The significance of digital gene expression profiles. *Genome Res.* 7:986-95.

Bhattacharya D, Aubry J, Twait EC, Jurk S. (2000) Actin gene duplication and the evolution of morphological complexity in land plants. *J. Phycol.* 36: 813-820.

Bonaldo, M. F., Lennon, G. and Soares, M. B. 1996. Normalization and subtraction: two approaches to facilitate gene discovery. *Genome Res.* 6:791-806.

Cole, K. M. and Conway, E. 1975. Phenetic implications of structural features of the perennating phase in the life history of *Porphyra* and *Bangia* (Bangiophyceae, Rhodophyta). *Phycologia* 16:205-16.

Endo H., Ootsuka S., Fukuda S., Kitade Y., Saga N. 2006. Functional complementation of an arginine auxotrophic yeast mutant by an argininosuccinate synthetase from *Porphyra yezoensis* (Rhodophyta). *J. Phycol.* 42: 1066-1071.

Kapraun, D. F., Hinson, T. K. and Lemus, A. J. 1991. Karyology and cytophotometric estimation of Inter- and Intraspecific nuclear DNA variation in four species of *Porphyra* (Rhodophyta). *Phycologia* 30:458-66.

Kitade, Y., Yamazaki, S. and Saga, N. 1996. A method for extraction of high molecular weight DNA from the macroalga *Porphyra yezoensis* (Rhodophyta). *J. Phycology* 32:496-8.

Kitade Y., Nakamura M., Endo H., Fukuda S., Kuwano K., Saga N. 2006. Characterization of a cDNA encoding a homolog of actin-related protein 4 from a marine red algae *Porphyra yezoensis*. *Fisheries Sci.* 72: 639-645.

Kitade Y., Nakamura M., Uji T., Fukuda S., Endo H., Saga N. 2008. Structural features and gene-expression profiles of actin homologs in *Porphyra yezoensis* (Rhodophyta). *Gene* 423: 79-84.

Kitade Y., Asamizu E., Fukuda S., Nakajima M., Ootsuka S., Endo H., Tabata S., Saga N. 2008. Identification of genes preferentially expressed during asexual sporulation in *Porphyra yezoensis* gametophytes (Bangiales, Rhodophyta). *J. Phycol.* 44: 113-123.

Kuwano, K., Aruga, Y. and Saga, N. 1996. Cryopreservation of clonal gametophytic thalli of *Porphyra* (Rhodophyta), *Plant Sci.* 116:117-24.

Le Gall, Y., Brown, S., Marie, D., Mejjad, M. and Kloareg, B. 1993. Quantification of nuclear DNA and GC content in marine macroalgae by flow cytometry of isolated nuclei. *Protoplasma* 173:123-32.

Nikaido, I., Asamizu, E., Nakajima, M., Nakamura, Y., Saga, N. and Tabata, S. 2000. Generation of 10,154 expressed sequence tags from a leafy gametophytes of a marine red alga, *Porphyra yezoensis*. *DNA Res.* 7:223-7.

Otero A., Vincenzini M. 2004. *Nostoc* (Cyanophyceae) goes nude: extracellular polysaccharides serve as a sink for reducing power under unbalanced C/N metabolism. *J. Phycol.* 40: 74-81.

Riley, M. 1993. Functions of the gene products of *Escherichia coli*, *Microbiol. Rev.* 57:862-952.

Schena M., Davis R.W. (2000) Technology standards for microarray research *In* Schena M. [Ed.] *Microarray Biochip Technology*. Eaton Publishing, Natick, Massachusetts, pp. 1-18.

In: Porphyra yezoensis
Editor: Koji Mikami

ISBN 978-1-61122-304-0

Chapter 6

TRANSIENT GENE EXPRESSION SYSTEMS IN *PORPHYRA YEZOENSIS*: ESTABLISHMENT, APPLICATION AND LIMITATION

***Koji Mikami*[*1] *and Toshiki Uji*[2]**

[1] Faculty of Fisheries Sciences, Hokkaido University, 3-1-1 Minato-cho, Hakodate 041-8611, Japan

[2] Graduate School of Fisheries Sciences, Hokkaido University, 3-1-1 Minato-cho, Hakodate 041-8611, Japan

ABSTRACT

Porphyra yezoensis has been recognized as a model for physiological and molecular biological studies of marine red macroalgae. However, due to the lack of transient and genetic transformation systems, little progress has been made in understanding the molecular mechanisms regulating the development and stress response. Recently, we, therefore, developed a transient gene expression system using the particle bombardment method in *P. yezoensis.* In this system, the β-glucuronidase (GUS) gene was shown to be useful through optimization of codon usage of the coding region to those of *P. yezoensis* nuclear genes and by employing a strong promoter of endogenous genes, namely, glyceraldehyde-3-phosphate dehydrogenase (GAPDH) and actin 1 (Act1) genes. It has been shown that the GUS reporter system is applicable for analysis of the promoter region of *P. yezoensis* nuclear genes. Moreover, transient gene expression with humanized fluorescent proteins was also established, enabling subcellular localization of phosphoinositides and transcription factors in *P. yezoensis* cells to be monitored. These applications will provide new opportunities for further investigations into the regulation and functions of various genes in *P. yezoensis* and other red macroalgae.

* Correspondence to: Koji Mikami, Faculty of Fisheries Sciences, Hokkaido University, 3-1-1 Minato-cho, Hakodate 041-8611, Japan, Tel/Fax: +81-138-40-8899, E-mail: komikami@fish.hokudai.ac.jp.

INTRODUCTION

Elucidation of the regulatory mechanisms by which plants respond and adapt to environmental stresses and control their development is essential for progress in fundamental plant biology and breeding (Krogan and Long, 2009; Manavalan et al., 2009). The molecular mechanisms regulating gene expression during stress responses and development have, therefore, been studied extensively in land plants and green algae, with genetic transformation being a key experimental system (Zalewski et al., 2010; Serrano et al., 2009).

To date, the function of genes critical for the regulation of stress responses and development in plants have been elucidated via genetic transformation systems. For example, in *Arabidopsis thaliana*, the genes encoding transcription factors were characterized as a key player in regulating the expression of a number of genes in response to drought, high salinity and low temperature (Nakashima et al., 2009), while the important roles of genes in leaf and flowering development were also revealed (Hake et al., 2004; Irish, 2010). In the moss *Physcomitrella patens*, a gene repressing the initiation of sporophyte and reproductive organ development has been identified (Okano et al., 2009), as well as MADS-box genes involved in gametangia formation and sporophyte production (Singer and Ashton, 2010). In addition, a kinesin gene from the unicellular green alga *Chlamydomonas reinhardtii* was shown to be a regulator of flagellar disassembly and flagellar assembly (Piao et al., 2009), and the role of a gene-encoding transcription factor regA from multicellular green alga *Volvox carteri* was further determined as a master regulator in differentiation of somatic and germ cells (Kirk and Nishii, 2001). Moreover, the gene encoding a central nitrogen assimilation regulator was identified in the unicellular red alga *Cyanidioschyzon merolae* (Imamura et al., 2009). In contrast, however, although a few genes involved in responses to environmental change and developmental programs have been identified in red macroalga (Kakinuma et al., 2006; 2008), functional analysis of these genes is impossible since genetic transformation as a tool for analyzing gene function has not yet been established in red macroalgae.

The marine red macroalga *Porphyra yezoensis* has several advantages for biological research; for example, it has a small genome size (2.6×10^8 bp; Matsuyama-Serisawa et al., 2007), a small number of chromosomes (n=3 in the haploid phase; Cole, 1990), a short life cycle (two to three months) in laboratory culture and a simple cultivation system (Saga and Kitade, 2002). In addition, analyses of expressed sequence tags (ESTs) from filamentous sporophytes and leafy gametophytes have already been performed in *P. yezoensis* (Nikaido, et al., 2000; Asamizu et al., 2003). Accordingly, *P. yezoensis* has attracted significant attention as a model marine plant for physiological and molecular biological studies of red macroalgae (Saga and Kitade, 2002; Waaland et al., 2004). However, the lack of transformation systems prevents progress in studies aimed at elucidation of the regulation and functions of protein-coding genes in *P. yezoensis*.

To resolve this, we recently developed reproducible and efficient transient gene expression systems in *P. yezoensis*, allowing us, for example, to identify transcription regulatory sequences in promoter regions and visualize subcellular localization of gene products in living cells. This chapter describes the development and successful application of the transient gene expression system in *P. yezoensis* with a brief summary of the current status of algal transformation. The problems creating disadvantage in these systems are also discussed.

Transformation Systems Established in Plants: A Brief Summary

Genetic transformation can be divided into stable and transient transformations under control of genes introduced into cells. In stable transformation, genes are maintained in the genome through generations of cells, whereas in transient transformation, rapid loss of introduced foreign genes is usually observed. Establishment of these transformation systems requires development of DNA transfection systems as well as identification of reporter genes, selection markers, suitable promoters to drive the reporters, and selection marker genes.

DNA Transfection Systems

Several methods of DNA transfection into cells are available in plants. Transformation mediated by the soil plant pathogen *Agrobacterium tumefaciens* has become the most used method for higher plant genetic transformation. With *A. tumefaciens*, it is possible to transfer a particular DNA segment, known as T-DNA, of a tumor-inducing plasmid into the nuclear genome of infected cells (Rao et al., 2009).

Particle bombardment, which is also referred to as micro-projectile bombardment, particle gun or biolistics, makes use of DNA-coated heavy-metal particles (mostly gold). This method enables the transient and stable transformation of almost any type of cell, regardless of rigidity of the cell wall, and is thus extensively used for land plants and algae such as the green alga *V. carteri* as well as with the moss *P. patens* and liverwort *Marchantia polymorpha* (Kianianmomeni et al., 2008; Cove, 2005; Takenaka et al., 2000). However, this method requires expensive devices and reagents such as a gene gun and gold particles.

Electroporation uses a high-voltage electrical pulse that temporarily disturbs the phospholipid bilayer of the plasma membrane, allowing cells to take up plasmid DNAs (Newell, 2000). This method is useful with *Arabidopsis* and tobacco protoplasts (Miao and Jiang, 2007). Another method, the polyethylene glycol (PEG)-mediated transformation system, makes use of PEG, which is also thought to affect the plasma membrane and induce uptake of DNAs into cells (Newell, 2000; Radchuk et al., 2002). PEG-mediated transformation is almost exclusively applied with the moss *P. patens* (Cove, 2005). Electroporation and PEG-mediated gene transfer are less expensive than particle bombardment in terms of the cost of both the apparatus and reagents, although both require culturing of protoplasts for regeneration of the plants (Newell, 2000; Radchuk et al., 2002).

The silicon carbide whisker-mediated transformation system is a simple method in which plasmid DNA is added to plant tissue and then mixed in a vortex (Frame et al., 1994), causing DNA-coated whiskers to penetrate the cell wall through small holes created by collisions between the plant cells and whiskers. This method has been employed to transform unicellular green alga and the callus of land plants (Dunahay, 1993; Asad et al., 2008). Similarly, the glass-bead abrasion technique is a simple method that involves agitation of cells in a vortex mixer in the presence of DNA and glass beads. Cells are physically damaged, allowing DNA to be introduced in the same way as with particle bombardment. This technique has been employed to transform unicellular green alga (Kindle, 1990; Feng et al., 2009).

Reporter Genes

Reporter genes are generally used to monitor whether genes of interest are expressed in cells by observing distinguishable expression over an endogenous cellular background. Such genes often code an enzyme that converts a substrate into a colored product or produces light emission. In some cases, a reporter gene product is itself a fluorescent protein. Reporter genes are usually employed by fusing them with promoters, allowing monitoring of gene expression and a coding region or protein domain during visualization of protein localization.

The *Escherichia coli uidA* encoding β-glucuronidase (GUS) is widely used as a reporter gene in land plants, because its expression patterns during the stress response and developmental process can be visualized as a blue color following treatment with X-gluc (5-bromo-4-chloro-3-indolyl-β-D-glucuronide). In addition, these patterns are then quantifiable by fluorometric analysis (Basu et al., 2004; Sun et al., 2010). Another color producing reporter is the *E. coli lacZ* encoding β-galactosidase, which cleaves the colorless substrate X-gal (5-bromo-4-chloro-3-indolyl-β-galactopyranoside) into galactose, resulting in a blue insoluble product (Bell et al., 2005). These color reporters are usually used to monitor gene expression *in planta*; however, visualization of the reporter products requires cell killing.

Reporters that function in living cells have also been established. For example, the bacterial luxAB system and luciferases from firefly and Renilla produce light emission in living cells (Greer and Szalay, 2002) and have successfully been applied to monitoring of gene expression and selection of mutants in which expression levels of the luciferase gene are altered (Dong et al., 2009). In addition, the green fluorescent protein (GFP) from jellyfish *Aequorea victoria*, reef coral *Zoanthus* sp. and copepoda *Pontellina plumata* has the advantage over other reporters for monitoring subcellular localization of proteins in living cells that its fluorescence can be visualized without additional substrates or cofactors (Ehrhardt, 2003). The availability of GFP variants with non-overlapping emission spectra such as cyan fluorescent protein (CFP), yellow fluorescent protein (YFP) and red fluorescent protein (RFP) makes multicolor imaging of cells *in vivo* feasible (Lin et al., 2008; Martin et al., 2009).

Selection Marker Genes

One of the functions of selection marker genes is to distinguish between transformed cells and non-transformed cells, since successful integration of a foreign gene into a host genome and expression of transferred DNA usually occur in only a small percentage of transfected cells. These genes confer new traits to any transformed target strain of a certain species, thus enabling the transformed cells to survive on medium containing the selective agent, where non-transformed cells die.

Antibiotic-resistance genes are generally used as selection markers. Aminoglycoside antibiotics, including neomycin, geneticin (G418), paromomycin and hygromycin B, are highly toxic to plant, animal and fungal cells as a result of binding to ribosomal subunits, which inhibits protein synthesis in bacteria, eukaryotic plastids and mitochondria (Miki and McHugh, 2004). It has been shown that neomycin phosphotransferase II (NPTII) detoxifies neomycin, G418 and paromomycin (Miki and McHugh, 2004), whereas hygromycin B phosphotransferase (HPT) inactivates hygromycin B via an ATP-dependent phosphorylation (Tian et al., 2000). Thus, bacterial NPTII and HPT genes are often used for mammalian, yeast and plant cells. Bleomycin is an antibiotic that belongs to the bleomycin family of glycopeptides, which act through site-specific, single- and double-stranded DNA cleavage (Perez et al., 1989). The bacterial *ble* gene conferring resistance to bleomycin has been used for green algae (Lumbreras et al., 1998; Hallmann and Rappel, 1999).

Since red algae usually present strong resistance to these antibiotics, there are currently no antibiotics-based selection markers for either unicellular or multicellular red algae.

Promoters

Promoters suitable for efficient expression of reporters and selection marker genes are very important transformation tools. The promoter of the cauliflower mosaic virus (CaMV) 35S RNA gene has been most frequently used as a constitutive strong promoter in plants (Yu et al., 2009; Kalamaki et al., 2009). In addition, both the rice actin (*Act1*) and maize ubiquitin (*Ubi1*) promoters have been shown to promote constitutive high level expression of foreign genes in monocotyledonous plants (McElroy et al., 1991; Katiyar-Agarwal et al., 2002). In the moss *P. patens*, the *CaMV 35S* promoter and rice *Act1* promoter have been tested using the GUS reporter system, indicating that the rice *Act1* promoter exhibits two- to tenfold higher activity than the *CaMV 35S* promoter in *P. patens* cells (Horstmann et al., 2004). Moreover, to avoid the toxicity of high level constitutive expression of reporter genes, stress- or steroid hormone-inducible promoters have been used for plant transformation (Masclaux et al., 2004; Ko et al., 2009). For example, a heat-inducible expression system has been developed using the *Arabidopsis* small heat shock protein 18.2 promoter in tobacco hairy root (Lee et al., 2007a), while a steroid hormone-inducible system in *Arabidopsis* plants has been established using a dexamethasone-inducible promoter (Aoyama and Chua, 1997).

Current Status of Algal Transformation

Within algae, transient and stable transformations have successfully been established only in microalgae as shown in Table 1 (Walker et al., 2005; Hallmann, 2007). Nuclear stable transformation of the unicellular green alga *C. reinhardtii*, a model for studying photosynthesis, chloroplast biogenesis and flagellar function (Rochaix, 1995), has been developed using various methods including particle bombardment (Debuchy et al., 1989), silicon carbide whiskers (Shimogawara et al., 1998), glass beads (Kindle, 1990) and electroporation (Shimogawara et al., 1998), all of which are based on complement auxotrophic mutants (Debuchy et al., 1989; Kindle et al., 1989). In *C. reinhardtii*, although endogeneous argininosuccinate lyase (ARG) and nitrate reductase (NIT) genes were originally used as selectable maker genes (Debuchy et al., 1989; Kindle et al., 1989), several selection markers based on bacterial antibiotic resistance genes are now available, including the bleomycin resistance gene *ble* (Lumbreras et al., 1998), paromomycin resistance gene *aphVIII* (Sizova et al., 2001) and hygromycin B resistance gene *aph7* (Berthold et al., 2002).

Table 1. Summary of stable and transient transformation in algae

Pylum Species	Transformation	Promoter	Selection marker Reporter	Reference
Microalgae				
Rhodophyta				
Cyanidioschyzon merolae	transient	β-tubulin	Hemagglutinin tag	Ohnuma et al., 2008
Cyanidioschyzon merolae	stable	URA5.3	URA5.3	Minoda et al., 2004
Porphyridium sp.	stable (chloroplast)		AHAS (W492S)	Lapidot et al., 2002
Chlorophyta				
Chlamydomonas reinhardtii	stable	β2-tubulin	aph7	Berthold et al., 2002
Chlamydomonas reinhardtii	stable (chloroplast)	RbcL/AtpA	aadA	Goldschmidt-Clermont, 1991
Volvox carteri	stable	Hsp70/RbcS	aphH	Jakobiak et al., 2004
Gonium pectorale	stable	VcHsp70/RbcS	aphVIII	Lerche and Hallmann, 2009
Dunaliella salina	stable	CrRbcS2	ble	Jin et al., 2001
Bacillariophyta				
Cyclotella cryptica	stable	nptII	CcAcc	Dunahay et al., 1995
Navicula saprophila	stable	nptII	CcAcc	Dunahay et al., 1995
Phaeodactylum tricornutum	stable	ble	FcpA	Zaslavskaia et al., 2000
Alveolata				
Amphidinium sp	stable	Nos	nptII	ten Lohuis and Miller, 1998
Macroalgae				
Rhodophyta				
Porphyra yezoensis	transient	GAPDH	PyGUS	Fukuda et al., 2008
Porphyra yezoensis	transient	Act1	AmCFP	Mikami et al., 2009
Porphyra yezoensis	transient	Act1	PyGUS	Takahashi et al., 2010
Porphyra yezoensis	transient	Act1	sGFP/ZsGFP/ZsYFP	Uji et al., 2010
Porphyra yezoensis	transient	β-tubulin	GUS*	Gong et al., 2005
Porphyra yezoensis	transient	CaMV 35S	GUS*	Kuang et al., 1998
Porphyra miniata	transient	CaMV 35S	GUS*	Kübler et al., 1994
Porphyra tenera	transient	CaMV 35S	GUS*	Okauchi and Mizukami, 1999
Kappaphycus alvarezii	transient	CaMV 35S	GUS*	Kurtzman and Cheney, 1991
Chlorophyta				
Ulva pertusa	transient	RbcS	EGFP	Kakinuma et al., 2009
Phaeophyta				
Laminaria japonica	transient	CaMV 35S	GUS	Li et al., 2009a

*No significant GUS enzymatic activity was observed.

Similary, several reporter genes have also been developed for *C. reinhardtii*, such as the sulfer starvation-inducible gene for arylsulfatase (ARS) (Davies et al., 1992; 1994), which is easily assayed using a chromogenic sulfate substrate and can be used to measure changes in promoter activity in response to various environmental cues (Kucho, et al., 2003; Fei et al., 2009). In addition, genes encoding GFP and luciferase adapted to *C. reinhardtii* codon usage have also been developed (Fuhrmann et al., 1999; 2004). Moreover, several endogenous promoters have proven to be of high value for efficient expression, such as the ribulose bisphosphate carboxylase, small chain (RbcS2) promoter (Stevens et al., 1996), heat shock protein 70A (*Hsp70A*)/*RbcS2* fusion promoter (Schroda et al, 2000) and *Hsp70A*/β2-tubulin (*β2-tub*) fusion promoter (Schroda et al, 2000). Furthermore, stable transformation of chloroplast genomes has already been developed using the bacterial gene *aadA* and particle bombardment method (Goldschmidt-Clermont, 1991).

Nuclear transformation systems have been also developed for green Volvocaceae algae, which are becoming increasingly popular as a developmental genetic and evolutionary model (Kirk and Nishii, 2001; Kirk, 2005). Stable transformation has been developed based on the bacterial *aphH* and *aphVIII* genes in multicellular *V. carteri* and *Gonium pectorale*, respectively (Jakobiak et al., 2004; Lerche and Hallmann, 2009), with the *Hsp70*/*RbsS* fusion promoter from *V. carteri*. Stable nuclear transformation of the halotolerant unicellular green alga *Dunaliella salina* has also been reported using the *ble* gene under control of the *C. reinhardtii RbcS2* promoter (Jin et al., 2001). Transient transformation in this species has been successful using the GUS reporter gene and particle bombarment method (Tan et al., 2005). In unicellular green alga *Chlorella* species, stable and trasient transformation have been also reported (Dawson et al., 1997; Chen et al., 2001). In contrast to green microalgae, however, only transient gene expression of the GFP reporter gene by directed endogeneous *RbcS* promoter has been reported in green macroalga (Kakinuma et al., 2009).

Stable transformation of the diatoms *Cyclotella cryptica* and *Navicula saprophila* has been established with the bacterial *nptII* gene driven by a promoter from the *C. cryptica* acetyl-CoA carboxylase (ACC) gene (Dunahay et al., 1995), and subsequently, stable transformation for the pennate diatom *Phaeodactylum tricornutum* has been developed (Zaslavskaia et al., 2000). In two dinoflagellate species, *Amphidinium* sp. and *Symbiodinium microadriaticum*, three different bacterial genes (*nptII*, *hpt*, and *GUS*) have been successfully expressed by delivering the plasmid DNA into cells by agitation in the presence of silicon carbide whiskers (ten Lohuis and Miller, 1998).

For unicellular red algae, nuclear transformation has been established only in *C. merolae* (Minoda et al., 2004; Imamura et al., 2009; 2010), which consists of a minimum set of organelles: one nucleus, one mitochondrion and one plastid (Matsuzaki et al., 2004) and is recognized as a model organism for the investigation of the basic architecture of eukaryotes, including organelle division machinery (Kuroiwa et al., 2006; Yoshida et al., 2009). Since stable transformation has been developed using a uracil auxotrophic mutant of *C. merolae* lacking the *URA5.3* gene, the wild-type *URA5.3* gene fragment has been used for isolation of mutants with uracil prototrophic (Minoda et al., 2004). Using this system, mutants lacking the gene encoding R2R3-type MYB transcription factor or nitrate reductase were obtained (Imamura et al., 2009; 2010). For transient transformation, the plasmid DNA containing the β-tubulin gene from *C. merolae* and a hemagglutinin (HA) tag was introduced into *C. merolae* cells using the PEG method, allowing visualization of intracellular localization of the tagged protein by anti-HA immunocytochemistry (Ohnuma et al., 2008). In addition, the GFP reporter has been developed as a selection marker for transient transformed cells (Ohnuma et al., 2009). Moreover, chloroplast transformation has been reported in the unicellular red alga *Porphyridium* sp. through integration of the gene encoding AHAS(W492S) into the chloroplast genome by homologous recombination, resulting in sulfometuron methyl (SMM) resistance at a high frequency in SMM-resistant colonies (Lapidot et al., 2002).

In contrast to microalgae, stable transformation has not yet been established in macroalgae (seaweed), although preliminary expreiments have been reported in *P. yezoensis* and the Phaeophyta *Laminaria japonica* (Cheney et al., 2001; Jiang et al., 2002). In addition, transient transformation by particle bombardment has been reported in only a few macroalgal species (Table 1) such as *Ulva pertusa* (Kakinuma et al., 2009) and *L. japonica* (Li et al., 2009a). However, several reports have documented transient transformation in red macroalgae in which the expression construct contains the GUS reporter gene from *E. coli* under the control of the *CaMV 35S* promoter (Table 1). For example, transient gene expression has been reported using the 35S-GUS fusion gene for *Porphyra* species through delivery into the cells by electroporation, PEG-mediated gene transfer and particle bombardment (Kübler et al., 1994; Kuang et al., 1998). Transient gene expression via particle bombardment has also been reported with 35S-GUS in the important red macroalga *Kappaphycus alvarezii* (Kurtzman and Cheney, 1991). However, these reports with 35S-GUS gene expression have problems in terms of repetition in other laboratories.

Development and Application of the GUS Reporter System in *P. yezoensis*

Establishment of the PyGUS Reporter System

As shown in Table 1, transient transformation of Bangiophycideae has previously been attempted using the CaMV 35S RNA promoter-GUS reporter gene; however, the GUS enzymatic activity presented in these previous studies shows basal background levels (Kübler et al., 1994; Kuang et al., 1998; Okauchi and Mizukami, 1999). Indeed, when p35S-GUS was bombarded into gametophytic cells of *P. yezoensis* using particle bombardment in our laboratory, no GUS expression was observed (Figure 1A).

Similar defects of transient gene expression have been observed in *C. reinhardtii*. Foreign genes are often expressed poorly in *C. reinhardtii* because of the high GC content in protein-coding regions of nuclear genes from this organism (Fuhrmann et al., 1999; Cerutti et al., 1997). However, optimization of codon usage of GFP and luciferase genes to reflect that of the *C. reinhardtii* nuclear genome has provided high-level expression of these modifed genes (Fuhrmann et al., 1999; 2004). Because the GC content of *P. yezoensis* nuclear genes inferred from EST analysis reached a high of 65.2% as with the *C. reinhardtii* nuclear genes, it is possible that optimization of codon usage of foreign genes such as the AT rich-GUS gene is important for efficient expression in *P. yezoensis* cells. As expected, optimization of codon usage of the GUS-coding region to the preferred triplets in *P. yezoensis* (designated PyGUS) increased the GC content from 51.8 to 66.6%, resulting in significant but very low-level expression of 35S-PyGUS in gametophytic cells (Figure 1A).

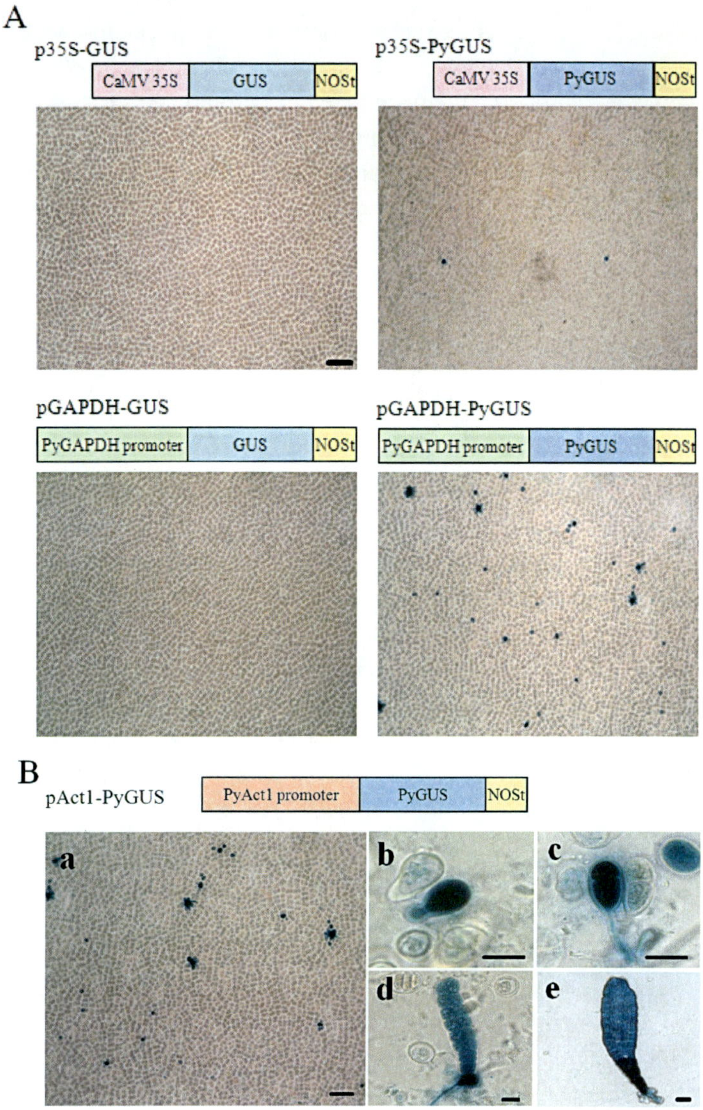

Figure 1 Expression of PyGUS under the direction of the *PyGAPDH* and *PyAct1* promoters in *P. yezoensis* cells. Expression of PyGUS under the direction of the *PyGAPDH* promoter in gametophytic cells of *P. yezoensis*. Structures of fusion genes used for each particle bombardment experiment are schematically represented above each photo. The 835 bp *CaMV 35S* (pink) and 2.2 kb *PyGAPDH* (yellow green) promoters were fused to native and synthetic GUS-coding regions (sky blue and blue, respectively), which resulted in the construction of 35S-PyGUS, PyGAPDH-GUS and PyGAPDH-PyGUS. NOSt (orange) indicates the *nos* gene terminator (253 bp). Gametophytic cells of *P. yezoensis* were transiently transformed with expression plasmids p35S-GUS, p35S-PyGUS, pGAPDH-GUS and pGAPDH-PyGUS by particle bombardment. Gametophytic thalli were then examined by microscopy 48 h after transient transformation. The scale bar corresponds to 100 μm. (B) Normal regeneration of transiently transformed protoplasts of *P. yezoensis* with expression plasmids pAct1-PyGUS. Structures of fusion genes used for the particle bombardment experiment are schematically represented in the top panel. The 3.0-kbp long *PyAct1* promoter (red) was fused to PyGUS (blue)-coding regions, resulting in the construction of pAct1-PyGUS. PyGUS expression was observed in particle-bombarded gametophytic cells (a) and during regeneration of these protoplasts (b-e). Scale bars correspond to 10 μm.

It is worth noting that attempts to express foreign genes under the control of the *CaMV 35S* promoter in *C. reinhardtii* have been unsuccessful (Day et al., 1990; Blankenship and Kindle, 1992); however, employment of endogenous promoters such as *β-tub*, *RbcS2* and *Hsp70* has allowed feasible foreign gene expression at a high level in *C. reinhardtii* (Davies et al., 1992; Stevens et al., 1996; Schroda et al., 2000). Similarly, the low efficient expression of 35S-PyGUS was improved using endogenous strong glyceraldehyde-3-phosphate dehydrogenase (GAPDH) and actin 1 (*PyAct1*) promoters instead of the *CaMV 35S* promoter (Figure 1).

These findings demonstrate that optimization of codon usage of the foreign gene and use of an endogenous strong promoter are crucial for achieving reproducible and efficient foreign gene expression in *P. yezoensis* (Fukuda et al., 2008).

Application of the PyGUS Reporter System for Promoter Analysis

A transient gene expression system has been successfully applied to promoter analyses in many kinds of organisms (Jin et al., 2009; Thonpho et al., 2010; Trong-Tue et al., 2010; Zhou et al., 2010). Because promoter analyses can give an accurate understanding of regulatory systems in gene expression, promoters of genes involved in environmental stress response and development have been extensively analyzed. In land plants, regulatory elements that respond, for example, to drought, cold stress, and pollen development have been identified through promoter analyses using the GUS reporter system (Xu et al., 2008; Chen et al., 2010). However, promoter analysis through transient gene expression has not yet been reported in macroalgae.

P. yezoensis undergoes a heteromorphic alternation of generations with a leafy gametophyte and a filamentous sporophyte, which presents many different cellular characteristics such as nuclear ploidy and cell wall composition (Cole and Conway, 1975; Mukai et al., 1981). To elucidate the molecular mechanisms underlying these differences, we analyzed the promoter region of a sporophyte-specific gene using the PyGUS reporter system, resulting in identification of the regulatory region of sporophyte-specific expression (Uji et al., unpublished data). To date, genes as candidates involved in the development and stress response have been identified using subtracted cDNA libraries in *P. yezoensis* (Kakinuma et al., 2006; 2008; Liu and Reith, 1996; Liu et al., 1996). Promoter analyses of these genes using the PyGUS reporter system will, therefore, contribute to elucidation of the molecular mechanisms directing the expression of temporally and environmentally regulated genes in *P. yezoensis*.

DEVELOPMENT AND APPLICATION OF THE FLUORESCENT REPORTER SYSTEM IN *P. YEZOENSIS*

In contrast to the GUS reporter system, analysis with fluorescent proteins such as GFP, CFP and YFP can be performed without any specific pretreatment in living cells by single and multicolor imaging in studies of protein localization (Nowak et al., 2004; Vermeer et al., 2009). Transient gene expression with fluorescent proteins has already been reported in microalgae. For example, functional analyses of plastid-targeting presequences of diatoms and Cryptophytes have been performed using GFP-presequence fusion (Apt et al., 2002; Kilian and Kroth, 2005; Gould et al., 2006). In *C. reinhardtii*, subcellular localization of the Rhesus protein (CrRh1) has been analyzed using CrRh1-GFP fusion, showing that CrRh1 localizes to the plasma membrane (Yoshihara et al., 2008). Furthermore, as mentioned above, GFP has also been used for selection of transiently transformed cells of *C. merolae*.

Until recently, there was no report of subcellular localization analysis using fluorescent proteins in macroalgae due to the lack of a transient gene expression system using fluorescent proteins; however, fluorescent reporter systems have recently been established in *P. yezoensis* (Mikami et al., 2009; Uji et al., 2010). Humanized fluorescent protein genes, AmCFP, ZsGFP, ZsYFP (Clontech), and plant-adapted GFP(S65T) (Niwa et al., 1999), the GC contents of which are as high as 63.7%, 62.8 %, 61.9% and 61.4%, respectively, were strongly expressed in gametophytic cells and monospores under the direction of the *PyAct1* promoter using the particle bombardment method (Figure 2). Thus, humanized and plant-adapted fluorescent reporters are useful for transient gene expression in *P. yezoensis*.

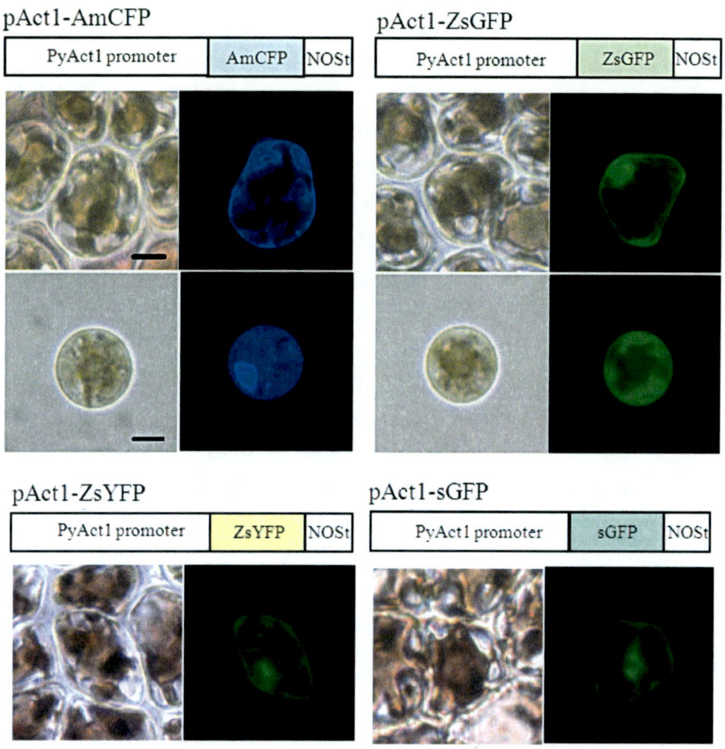

Figure 2 Expression of AmCFP, ZsGFP, ZsYFP and sGFP(S65T) in gametophytic cells and monospores of *P. yezoensis*. Structures of fusion genes used for each particle bombardment experiment are schematically represented above each photo. The 3.0-kbp long *PyAct1* promoter was fused to AmCFP (sky blue), ZsGFP (yellow green), ZsYFP (yellow) and sGFP (green)-coding regions, resulting in the construction of pAct1-AmCFP, pAct1-ZsGFP, pAct1-ZsYFP and pAct1-sGFP, respectively. Gametophytic cells and monospores of *P. yezoensis* were transiently transformed with expression plasmids pAct1-AmCFP, pAct1-ZsGFP, pAct1-ZsYFP and pAct1-sGFP by particle bombardment. Cells were then examined by fluorescent microscopy 48 h after transient transformation. Left and right panels show bright field and fluorescence images, respectively. The scale bar corresponds to 5 μm.

Visualization of Phosphoinositides

Despite the minority of membrane phospholipids, phosphoinositides (PIs) play important regulatory roles in eukaryotic stress response and development (Xue et al., 2009; Helmann, 2009). Phosphorylation of hydroxyl groups at positions D3, D4 and D5 in the inositol ring of phosphatidylinositol (PtdIns) gives rise to a family of structurally related lipids; PtdIns-monophosphates [PtdIns3P, PtdIns4P and PtdIns5P], PtdIns-bisphosphates [PtdIns(3,4)P_2, PtdIns(3,5)P_2 and PtdIns(4,5)P_2] and PtdIns-trisphosphate [PtdIns(3,4,5)P_3], and these PIs are detectable in plants, except for PtdIns(3,4,5)P_3 (Xue et al., 2009; Helmann, 2009). It is well known that the PI system is involved in regulating multiple processes of development and cell responses to environmental stimuli in land plants and green algae (Helmann, 2009; Williams et al., 2005). In red algae, Li et al. (2008; 2009b) demonstrated that PIs are involved in the establishment of cell polarity in *P. yezoensis* monospores; however, the presence of PIs in *P. yezoensis* cells has not been directly shown.

In model animal, plant and lower eukaryotic species, a powerful technique to monitor PIs *in vivo* through expression of a fluorescent protein fused to the Pleckstrin homology (PH) domain as a PI binding module has been developed (Vermeer et al., 2009; Szentpetery et al., 2009; Loovers et al., 2006). Based on a variety of substrate specificities, the PH domains from human phospholipase Cδ1 (PLCδ1) are employed for detection of PtdIns(4,5)P_2 (Lee et al., 2007b), whereas that from the v-akt murine thymoma viral oncogene homolog 1 (Akt1) has dual specificity in detection of both PtdIns(3,4)P_2 and PtdIns(3,4,5)P_3 (Nishio et al., 2007). These systems have been successfully applied in land plant cells. For example, Dowd et al. (2006) showed that GFP-PLCδ1PH localizes at the plasma membrane of growing pollen tips of *Petunia inflata*. In addition, van Leeuwen et al. (2007) showed that YFP-PLCδ1PH is concentrated at the plasma membrane in *Arabidopsis* roots in response to salt stress or during the final stage of tobacco cell division.

We employed PH domain-fluorescent proteins to monitor the presence of PIs in the plasma membrane of *P. yezoensis* cells (Mikami et al., 2009). As a result, visualization of PtdIns(3,4)P_2 and PtdIns(4,5)P_2 at the plasma membrane was successfully performed with humanized AmCFP and ZsGFP fused to the PH domains from PLCδ1 and Akt1 (Figure 3A). It is, therefore, possible to use these fusion proteins as a plasma membrane marker in *P. yezoensis* and other red macroalgae.

Visualization of Nuclear Localization of Transcription Factors

Transcription factors regulate gene expression levels by binding to specific DNA sequences, so-called *cis*-acting elements, present in the promoter region of protein-coding genes (Riaño-Pachón et al., 2008). Studies on algal transcription factors indicate important roles in stress response and development. For instance, the *regA* gene from *V. carteri* plays a central role as a master regulatory gene that encodes a transcription factor involved in the differentiation of germ and somatic cells through suppression of reproductive activities in somatic cells (Kirk and Nishii, 2001). *C. merolae* R2R3-type MYB transcription factor, CmMYB1, is a central nitrogen assimilation regulator (Imamura et al., 2009). In addition, fluorescent proteins have generally been used for analysis of subcellular localization of transcription factors.

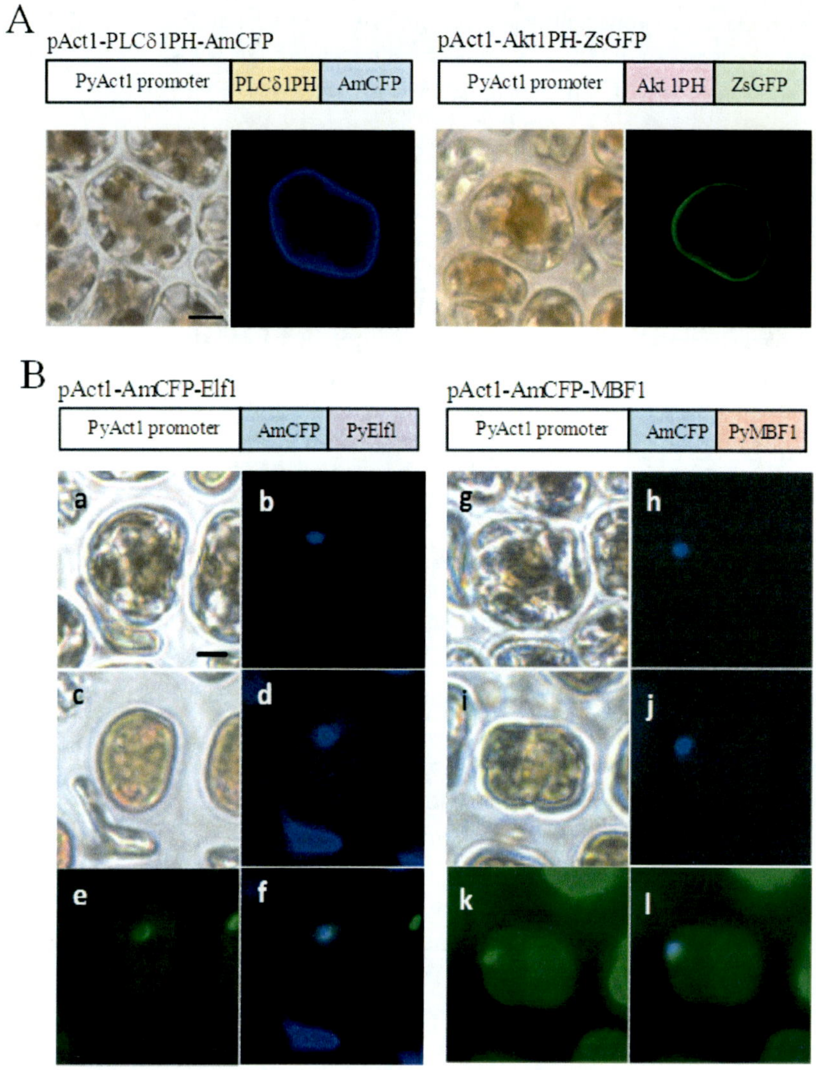

Figure 3. Visualization of subcellular localization with humanized fluorescent proteins in *P. yezoensis* cells. (A) Expression of the PLCδ1PH and Akt1PH domains fused with fluorescent proteins in gametophytic cells of *P. yezoensis*. Structures of fusion genes used for each particle bombardment experiment are schematically represented above each photo. The lengths of PLCδ1PH (orange) and Akt1PH (pink) are 510 bp and 607 bp, respectively. Expression plasmids pAct1-PLCδ1PH-AmCFP and pAct1-Akt1PH-ZsGFP were introduced into gametophytic cells of *P. yezoensis* by particle bombardment. Gametophytes were examined by fluorescent microscopy after 48 h of transient transformation. Left and right panels show bright field and fluorescence images, respectively. The scale bar corresponds to 5 μm. (B) Nuclear localization of PyElf1 and PyMBF1 in gametophytic cells of *P. yezoensis*. Structures of fusion genes used for each particle bombardment experiment are schematically represented above each photo. The lengths of ORFs encoding PyElf1 (purple) and PyMBF1 (red) are 255 bp and 384 bp, respectively. Gametophytic cells transiently transformed with expression plasmids pAct1-AmCFP-Elf1 (a to f) and pAct1-AmCFP-MBF1 (g to l). Gametophytes were examined by fluorescence microscopy after 48 h of transient transformation. Signals from AmCFP fusions (d and j) and SYBR Gold staining (e and k) were fixed with 80% methanol then stained with SYBR Gold (c-f and i-l). (f and l) represent merged images of AmCFP and SYBR Gold. Fluorescent images (b, d, e, f, h, j, k and l) are shown together with bright field images (a, c, g and i). The scale bar corresponds to 5 μm

For example, nuclear localization of asymmetric leaves 1 and 2 (AS1 and AS2) from *A. thaliana*, and rice ABA responsive element-binding factor 1 (OsABF1) were confirmed by transient gene expression as fusion proteins with GFP, CFP or YFP (Zhu et al., 2008; Hossain et al., 2010).

Subcellular localization of transcription factors from *P. yezoensis* has recently been analyzed using a transient gene expression system with fluorescent proteins (Uji et al., 2010). When complete, ORFs of transcription factors, transcription elongation factor 1 (PyElf1) and multiprotein bridging factor 1 (PyMBF1) from *P. yezoensis*, were fused to AmCFP or ZsGFP; nuclear localization of these fusion proteins was observed in gametophytic cells and confirmed by overlapping of fluorescent signals with SYBR Gold staining (Figure 3B).

Problems to be Addressed in the Near Future

It is clear that humanized and plant-adapted fluorescent proteins are employable as visual reporters in *P. yezoensis* to monitor subcellular localization of desired proteins (Figure 3). However, we observed that overexpression of fluorescent proteins had an inhibitory effect on the development of monospores (Mikami et al., 2009; Uji et al., 2010). Particle bombardment itself does not appear to be the cause of this inhibitory effect, because PyGUS-overexpressing monospores developed normally (Figure 1B) (Takahashi et al., 2010). It is, therefore, necessary to remove the inhibitory effects of overexpressed fluorescent protein to study the function of genes involved in the stress response and development of *P. yezoensis*.

In addition, visualization of protein localization in chloroplasts, mitochondria, and peroxysomes in *P. yezoensis* has yet to be achieved. In contrast, functional analyses of chloroplast-targeting sequences have been performed by transient gene expression using fluorescent protein in diatoms, dinoflagellates and Cryptophytes (Kilian and Kroth, 2005; Gruber et al., 2007), demonstrating similarities between the plastid protein import machineries in diatoms and Cryptophytes. Thus, establishment of visualization systems for protein localization in chloroplasts, mitochondria and peroxysomes is also necessary to understand the biological function of nuclear genes encoding organelle-targeting proteins in *P. yezoensis*.

Proposed Application of Transient Gene Expression Systems in Gene Silencing

Transient gene expression systems developed in *P. yezoensis* cells will contribute to functional analysis of genes using gene knock-down methods such as gene silencing using RNA interference (RNAi) and antisense RNA repression methods. RNAi, also known as posttranscriptional gene silencing (PTGS) in plants, involves mRNA degradation of the corresponding endogenous transcript through sequence-specific nucleotide interactions induced by double-stranded RNA (dsRNA) (Eamens et al., 2008). Gene knock-down approaches that trigger RNA silencing via transient expression of antisense or inverted repeat RNA have been successful in land plants and unicellular algae (Endo et al., 2008; Sun et al., 2008). In unicellular algae, gene silencing via the transient expression of antisense RNA has recently been successful for genes encoding catalase and the coiled-coil protein VIG1, which is essential for tethering vacuoles to mitochondria, in *C. merolae* (Ohnuma et al., 2009; Fujiwara et al., 2010). Moreover, gene knock-down via RNAi has also been achieved for phytoene desaturase, a key chloroplast enzyme in carotenoid biosynthesis in *D. salina* (Sun et al., 2008).

At present, gene silencing via transient expression has not yet been reported in macroalgae. As mentioned above, since overexpression of fluorescent proteins has an inhibitory effect on cell growth; no suitable markers currently exist for antisense RNA suppression and RNAi in red macroalgae. Again, removing these inhibitory effects from overexpressed fluorescent protein genes is necessary for the development of a gene-silencing system in red macroalgae.

CONCLUSION

Progress in molecular biological studies of red macroalgae has been slower than that of land plants and green algae due to the lack of transient and stable transformation systems. Recently, we have developed transient gene expression systems using the PyGUS gene as a reporter in *P. yezoensis*. There are two critical factors required for reproducibility and efficiency of transient gene expression, namely, optimization of codon usage of coding regions to those of *P. yezoensis* genes and employment of endogenous strong promoters. It has already been demonstrated that the PyGUS reporter system is useful for promoter analyses of *P. yezoensis* genes. These findings could be informative for the establishment of transient gene expression systems in other red macroalgae. Moreover, a transient gene expression system using humanized fluorescent proteins has been applied for visualization of subcellular localization of PIs and desired gene products in *P. yezoensis*, and could also provide new opportunities for analysis of subcellular localization of a variety of proteins from other red algae. Therefore, these transient gene expression systems will provide a new trustworthy tool for studying the regulation and functions of various genes in *P. yezoensis*.

The next step in improving related molecular biological techniques is to decrease the toxic effects of overexpressed fluorescent proteins and to establish a stable transformation system in *P. yezoensis*. As mentioned above, the former is necessary for selection markers of RNAi and antisense RNA supression to allow identification of transiently transformed cells. In addition, stable transformation systems are important for genetic analysis of gene function and molecular breeding of *P. yezoensis*. These improvements will enhance both our biological understanding and genetical engineering of *P. yezoensis* as a model for marine red macroalgae.

ACKNOWLEDGMENTS

This work was supported by a grant from the Hokusui Society Foundation.

REFERENCES

Aoyama, T. and Chua, N. H. (1997). A glucocorticoid-mediated transcriptional induction system in transgenic plants. *Plant Journal*, *11*, 605-612.

Apt, K. E., Zaslavkaia, L., Lippmeier, J. C., Lang, M., Kilian, O., Wetherbee, R., Grossman, A. R. and Kroth, P. G. (2002). *In vivo* characterization of diatom multipartite plastid targeting signals. *Journal of Cell Science*, *115*, 4061-4069.

Asad, S., Mukhtar, Z., Nazir, F., Hashmi, J. A., Mansoor, S., Zafar, Y. and Arshad, M. (2008). Silicon carbide whisker-mediated embryogenic callus transformation of cotton (*Gossypium hirsutum* L.) and regeneration of salt tolerant plants. *Molecular Biotechnology*, *40*, 161-169.

Asamizu, E., Nakajima, M., Kitade, Y., Saga, N., Nakamura, Y. and Tabata, S. (2003). Comparison of RNA expression profiles between the two generations of *Porphyra yezoensis* (Rhodophyta), based on expressed sequence tag frequency analysis. *Journal of Phycology*, *39*, 923-930.

Basu, C., Kausch, A. P. and Chandlee, J. M. (2004). Use of β-glucuronidase reporter gene for gene expression analysis in turfgrasses. *Biochemical and Biophysical Research Communications*, *320*, 7-10.

Bell, P., Limberis, M., Gao, G. P., Wu, D., Bove, M. S., Sanmiguel, J. C. and Wilson, J. M. (2005). An optimized protocol for detection of *E. coli* beta-galactosidase in lung tissue following gene transfer. *Histochemistry and Cell Biology*, *124*, 77-85.

Berthold, P., Schmitt, R. and Mages, W. (2002). An engineered S*treptomyces hygroscopicus aph 7*" gene mediates dominant resistance against hygromycin B in *Chlamydomonas reinhardtii*. *Protist*, *153*, 401-412.

Blankenship, J. E. and Kindle, K. (1992). Expression of chimeric genes by the light-regulated *cabII-1* promoter in *Chlamydomonas reinhardtii*: a *cabII-1/nit1* gene functions as a dominant selectable marker in a *nit1- nit2*-strain. *Mollecular Cell Biology*, *99*, 2074-2081.

Cerutti, H., Johnson, A. M., Gillham, N. W. and Boynton, J. E. (1997). A eubacterial gene conferring spectinomycin resistance on *Chlamydomonas reinhardtii*: Integration into the nuclear genome and gene expression. *Genetics*, *145*, 97-110.

Chen, L., Tu, Z. M., Hussain, J., Cong, L., Yan, Y. J., Jin, L., Yang, G. X. and He, G. Y. (2010). Isolation and heterologous transformation analysis of a pollen-specific promoter from wheat (*Triticum aestivum* L.). *Molecular Biology Reports*, *37*, 737-744.

Chen, Y., Wang, Y. Q., Sun, Y. R., Zhang, L. M. and Li, W. B. (2001). Highly efficient expression of rabbit neutrophil peptide-1 gene in *Chlorella ellipsoidea* cells. *Current Genetics*, *39*, 365-370.

Cheney, D. P., Metz, B. and Stiller, J. (2001). Agrobacterium-mediated genetic transformation in the macroscopic red alga *Porphyra yezoensis*. *Journal of Phycolgy*, *37* (Supplement) 11.

Cole, K. M. (1990). Chromosomes, In: Cole, K. M. and Sheath, R. G. (Eds). *Biology of the red algae*. (pp. 73-101). Cambridge, Cambridge University Press.

Cole, K. M. and Conway, E. (1975). Phenetic implications of structural features of the perennating phase in the life history of *Porphyra* and *Bangia* (Bangiophyceae, Rhodophyta). *Phycologia*, *16*, 205-216.

Cove, D. (2005). The moss *Physcomitrella patens*. *Annual Review of Genetics*, *39*, 339-358.

Davies, J. P., Weeks, D. P. and Grossman, A. R. (1992) Expression of the arylsulfatase gene from the beta 2-tubulin promoter in *Chlamydomonas reinhardtii*. *Nucleic Acids Research*, *20*, 2959-2965.

Davies, J. P., Yildiz, F. and Grossman, A. R. (1994). Mutants of *Chlamydomonas* with aberrant responses to sulfur deprivation. *Plant Cell*, *6*, 53-63.

Dawson, H. N., Burlingame, R. and Cannons, A. C. (1997). Stable transformation of *Chlorella*: Rescue of nitrate reductase-deficient mutants with the nitrate reductase gene. *Current Microbiology*, *35*, 356-362.

Day, A., Debuchy, R., Dillewijn, J., Purton, S. and Rochaix, J. D. (1990). Studies on the maintenance and expression of cloned DNA fragments in the nuclear genome of the green alga *Chlamydomonas reinhardtii*. *Physiologia Plantarum*, *78*, 254-260.

Debuchy, R., Purton, S. and Rochaix, J. D. (1989). The argininosuccinate lyase gene of *Chlamydomonas reinhardtii*: an important tool for nuclear transformation and for correlating the genetic and molecular maps of the *ARG7* locus. *EMBO Journal*, *8*, 2803-2809.

Dong, C. H., Zolman, B. K., Bartel, B., Lee, B., Stevenson, B., Agarwal, M. and Zhu, J. K. (2009). Disruption of *Arabidopsis* CHY1 reveals an important role of metabolic status in plant cold stress signaling. *Molecular Plant*, *2*, 59-72.

Dowd, P. E., Coursol, S., Skirpan, A. L., Kao, T. H. and Gilroy, S. (2006). Petunia phospholipase c1 is involved in pollen tube growth. *Plant Cell*. *18*, 1438-1453.

Dunahay, T. G. (1993). Transformation of *Chlamydomonas reinhardtii* with silicon carbide whiskers. *Biotechniques*, *15*, 452-455.

Dunahay, T. G., Jarvis, E. E. and Roessler, P. G. (1995). Genetic transformation of the diatoms *Cyclotella cryptica* and *Navicula saprophila*. *Journal of Phycology*, *31*, 1004-1012.

Eamens, A., Wang, M. B., Smith, N. A. and Waterhouse, P. M. (2008). RNA silencing in plants: Yesterday, today, and tomorrow. *Plant Physiology*, *147*, 456-468.

Ehrhardt, D. (2003). GFP technology for live cell imaging. *Current Opinion in Plant Biology*, *6*, 622-628.

Endo, S., Pesquet, E., Tashiro, G., Kuriyama, H., Goffner, D., Fukuda, H. and Demura, T. (2008). Transient transformation and RNA silencing in *Zinnia* tracheary element differentiating cell cultures. *Plant Journal*, *53*, 864-875.

Fei, X. W., Eriksson, M., Yang, J. H. and Deng, X. D. (2009). An Fe deficiency responsive element with a core Sequence of TGGCA regulates the expression of FEA1 in *Chlamydomonas reinharditii*. *Journal of Biochemistry*, *146*, 157-166.

Feng, S. Y., Xue, L. X., Liu, H. T. and Lu, P. J. (2009). Improvement of efficiency of genetic transformation for *Dunaliella salina* by glass beads method. *Molecular Biology Reports*, *36*, 1433-1439.

Frame, B. R., Drayton, P. R., Bagnell, S. V., Lewnau, C. J., Bullock, W. P., Wilson, H M., Dunwell , J. M., Thompson, J. A. and Wang, K. (1994). Production of fertile transgenic maize plants by silicon carbide whisker-mediated transformation. *Plant Journal*, *6*, 941-948.

Fuhrmann, M., Hausherr, A., Ferbitz, L., Schödl, T., Heitzer, M. and Hegemann, P. (2004). Monitoring dynamic expression of nuclear genes in *Chlamydomonas reinhardtii* by using a synthetic luciferase reporter gene. *Plant Molecular Biology*, *55*, 869-881

Fuhrmann, M., Oertel, W. and Hegemann, P. (1999). A synthetic gene coding for the green fluorescent protein (GFP) is a versatile reporter in *Chlamydomonas reinhardtii*. *Plant Journal*, *19*, 353-361.

Fujiwara, T., Kuroiwa, H., Yagisawa, F., Ohnuma, M., Yoshida, Y., Yoshida, M., Nishida, K., Misumi, O., Watanabe, S., Tanaka, K. and Kuroiwa, T. (2010). The Coiled-Coil Protein VIG1 Is Essential for Tethering Vacuoles to Mitochondria during Vacuole Inheritance of *Cyanidioschyzon merolae*. *Plant Cell*, *22*, 772-781.

Fukuda, S., Mikami, K., Uji, T., Park, E. J., Ohba, T., Asada, K., Kitade, Y., Endo, H., Kato, I. and Saga, N. (2008). Factors influencing efficiency of transient gene expression in the red macrophyte *Porphyra yezoensis*. *Plant Science*, *174*, 329-339.

Goldschmidt-Clermont, M. (1991). Transgenic expression of aminoglycoside adenine transferase in the chloroplast: a selectable marker for site-directed transformation of *Chlamydomonas*. *Nucleic Acids Research*, *19*, 4083-4089.

Gong, Q., Yu, W., Dai, J., Liu, H., Xu, R., Guan, H. and Pan, K. (2005). Efficient *gusA* transient expression in *Porphyra yezoensis* protoplasts mediated by endogenous beta-tubulin flanking sequences. *Journal of Ocean University of China*, *6*, 21-25.

Gould, S. B., Sommer, M. S., Hadfi, K., Zauner, S., Kroth, P. G. and Maier, U. G. (2006). Protein targeting into the complex plastid of cryptophytes. *Journal of Molecular Evolution*, *62*, 674-681.

Greer, L. F. and Szalay, A. A. (2002). Imaging of light emission from the expression of luciferases in living cells and organisms: a review. *Luminescence*, *17*, 43-74.

Gruber, A., Vugrinec, S., Hempel, F., Gould, S. B., Maier, U. G. and Kroth, P. G. (2007). Protein targeting into complex diatom plastids: functional characterization of a specific targeting motif. *Plant Molecular Biology*, *64*, 519-530.

Hake, S., Smith, H. M., Holtan, H., Magnani, E., Mele, G. and Ramirez, J. (2004). The role of knox genes in plant development. *Annual Review of Cell and Developmental Biology*, *20*, 125-151.

Hallmann, A. (2007). Algal Transgenics and Biotechnology. In: Teixeira da Silva J. A. and Shima, K (Eds). *Transgenic Plant Journal*. (pp 81-98). U. K. Global Science Books.

Hallmann, A. and Rappel, A. (1999). Genetic engineering of the multicellular green alga *Volvox*: a modified and multiplied bacterial antibiotic resistance gene as a dominant selectable marker. *Plant Journal*, *17*, 99-109.

Helmann, I. (2009). Using genetic tools to understand plant phosphoinositide s signalling. *Trends in Plant Science*, *14*, 171-179.

Horstmann, V., Huether, C. M., Jost, W., Reski, R. and Decker, E. L. (2004). Quantitative promoter analysis in *Physcomitrella patens*: a set of plant vectors activating gene expression within three orders of magnitude. *BMC Biotechnology*, *4*, 13. online.

Hossain, M. A., Lee, Y., Cho, J. I., Ahn, C. H., Lee, S. K., Jeon, J. S., Kang, H., Lee, C. H., An, G. and Park, P. B. (2010). The bZIP transcription factor OsABF1 is an ABA responsive element binding factor that enhances abiotic stress signaling in rice. *Plant Molecular Biology*, *72*, 557-566.

Imamura, S., Kanesaki, Y., Ohnuma, M., Inouye, T., Sekine, Y., Fujiwara, T., Kuroiwa, T. and Tanaka, K. (2009). R2R3-type MYB transcription factor, CmMYB1, is a central nitrogen assimilation regulator in *Cyanidioschyzon merolae*. *Proceedings of the National Academy of Sciences of the United States of America*, *106*, 12548-12553.

Imamura, S., Terashita, M., Ohnuma, M., Maruyama, S., Minoda, A., Weber, A. P. M., Inouye, T., Sekine, Y., Fujita, Y., Omata, T. and Tanaka, K. (2010). Nitrate assimilatory genes and their transcriptional regulation in a unicellular red alga *Cyanidioschyzon merolae*: Genetic evidence for nitrite reduction by a sulfite reductase-like enzyme. *Plant and Cell Physiology*, *51*, 707-717.

Irish, V. F. (2010). The flowering of Arabidopsis flower development. *Plant Journal*, *61*, 1014-1028.

Jakobiak, T., Mages, W., Scharf, B., Babinger, P., Stark, K. and Schmitt, R. (2004). The bacterial paromomycin resistance gene, *aphH*, as a dominant selectable marker in *Volvox carteri*. *Protist*, *155*, 381-393.

Jiang, P., Qin, S. and Tseng, C. K. (2002). Expression of hepatitis B surface antigen gene (HBsAg) in *Laminaria japonica* (Laminariales, Phaeophyta). *Chinese Science Bulletin*, *47*, 1438-1440.

Jin, D. Q., Ni, T. T., Hou, J., Rellinger, E. and Zhong, T. P. (2009). Promoter Analysis of *Ventricular Myosin Heavy Chain* (*vmhc*) in Zebrafish Embryos. *Developmental Dynamics*, *238*, 1760-1767.

Jin, E. S., Polle, J. E. W. and Melis, A. (2001). Involvement of zeaxanthin and of the Cbr protein in the repair of photosystem II from photoinhibition in the green alga *Dunaliella salina*. *Biochimica et Biophysica Acta-Bioenergetics 1506*, 244-259.

Kakinuma, M., Coury, D., Nakamoto, C., Sakaguchi, K. and Amano, H. (2008). Molecular analysis of physiological responses to changes in nitrogen in a marine macroalga, *Porphyra yezoensis* (Rhodophyta). *Cell Biology and Toxicology*, *24*, 629-639.

Kakinuma, M., Ikeda, M., Coury, D. A., Tominaga, H., Kobayashi, I. and Amano, H. (2009). Isolation and characterization of the *rbcS* genes from a sterile mutant of *Ulva pertusa* (Ulvales, Chlorophyta) and transient gene expression using the *rbcS* gene promoter. *Fisheries Science*, *75*, 1015-1028.

Kakinuma, M., Kaneko, I., Coury, D. A., Suzuki, T. and Amano, H. (2006). Isolation and identification of gametogenesis-related genes in *Porphyra yezoensis* (Rhodophyta) using subtracted cDNA libraries. *Journal of Applied Phycology*, *18*, 489-496.

Kalamaki, M. S., Alexandrou, D., Lazari, D., Merkouropoulos, G., Fotopoulos, V., Pateraki, I., Aggelis, A., Carrillo-Lopez, A., Rubio-Cabetas, M. J. and Kanellis, A. K. (2009). Over-expression of a tomato N-acetyl-L-glutamate synthase gene (SlNAGS1) in *Arabidopsis thaliana* results in high ornithine levels and increased tolerance in salt and drought stresses. *Journal of Experimental Botany*, *60*, 1859-1871.

Katiyar-Agarwal, S., Kapoor, A. and Grover, A. (2002). Binary cloning vectors for efficient genetic transformation of rice. *Current Science*, *82*, 873-876.

Kianianmomeni, A., Nematollahi, G. and Hallmann, A. (2008). A gender-specific retinoblastoma-related protein in *Volvox carteri* implies a role for the retinoblastoma protein family in sexual development. *Plant Cell*, *20*, 2399-2419.

Kilian, O. and Kroth, P. G. (2005). Identification and characterization of a new conserved motif within the presequence of proteins targeted into complex diatom plastids. *Plant Journal*, *41*, 175-183.

Kindle, K. L. (1990). High-frequency nuclear transformation of *Chlamydomonas reinhardtii*. *Proceedings of the National Academy of Sciences of the United States of America*, *87*, 1228-1232.

Kindle, K. L., Schnell, R. A., Fernandez, E. and Lefebvre, P. A. (1989). Stable nuclear transformation of *Chlamydomonas* using the *Chlamydomonas* gene for nitrate reductase. *Journal of Cell Biology*, *109*, 2589-2601.

Kirk, D. L. (2005). A twelve-step program for evolving multicellularity and a division of labor. *Bioessays*, *27*, 299-310.

Kirk, D. L. and Nishii, I. (2001). *Volvox carteri* as a model for studying the genetic and cytological control of morphogenesis. *Development Growth and Differentiation*, *43*, 621-631.

Ko, J. H., Kim, W. C. and Han, K. H. (2009). Ectopic expression of MYB46 identifies transcriptional regulatory genes involved in secondary wall biosynthesis in *Arabidopsis*. *Plant Journal*, *60*, 649-665.

Krogan, N. T. and Long, J. A. (2009). Why so repressed? Turning off transcription during plant growth and development. *Current Opinion in Plant Biology*, *12*, 628-636.

Kuang, M., Wang, S. J., Li, Y., Shen, D. L. and Zeng, C. K. (1998) Transient expression of exogenous GUS gene in *Porphyra yezoensis* (Rhodophyta). *Chinese Journal of Oceanology and Limnology*, *16* (Supplement) 56-61.

Kübler, J. E., Minocha, S. C. and Mathieson, A. C. (1994) Transient expression of the GUS reporter gene in protoplasts of *Porphyra miniata* (Rhodophyta). *Journal Marine Biotechnology*, *1*, 165-169.

Kucho, K., Yoshioka, S., Taniguchi, F., Ohyama, K. and Fukuzawa, H. (2003). Cis-acting elements and DNA-binding proteins involved in CO_2-responsive transcriptional activation of *Cah1* encoding a periplasmic carbonic anhydrase *Chlamydomonas reinhardtii*. *Plant Physiology*, *133*, 783-793.

Kuroiwa, T., Nishida, K., Yoshida, Y., Fujiwara, T., Mori, T., Kuroiwa, H. and Misumi, O. (2006). Structure, function and evolution of the mitochondrial division apparatus. *Biochimica et Biophysica Acta-Molecular Cell Research*, *1763*, 510-521.

Kurtzman, A. M. and Cheney, D. P. (1991). Direct gene transfer and transient expression in a marine red alga using the biolistic method. *Journal of Phycology*, *27* (Supplement) 42.

Lapidot, M., Raveh, D., Sivan, A., Arad, S. and Shapira, M. (2002). Stable chloroplast transformation of the unicellular red alga *Porphyridium* species. *Plant Physiology*, *129*, 7-12.

Lee, K. T., Chen, S. C., Chiang, B. L. and Yamakawa, T. (2007a). Heat-inducible production of beta-glucuronidase in tobacco hairy root cultures. *Applied Microbiology and Biotechnology*, *73*, 1047-1053.

Lee, Y., Kim, Y. W., Jeon, B. W., Park, K. Y., Suh, S. J., Seo, J., Kwak, J. M., Martinoia, E. and Hwang, I. (2007b). Phosphatidylinositol 4,5-bisphosphate is important for stomatal opening. *Plant Journal*, *52*, 803-816.

Lerche, K. and Hallmann, A. (2009). Stable nuclear transformation of *Gonium pectorale*. *BMC Biotechnology*, 9.

Li, F. C., Qin, S., Jiang, P., Wu, Y. and Zhang, W. (2009a). The integrative expression of GUS gene driven by *FCP* promoter in the seaweed *Laminaria japonica* (Phaeophyta). *Journal of Applied Phycology*, *21*, 287-293.

Li, L., Saga, N. and Mikami, K. (2008). Phosphatidylinositol 3-kinase activity and asymmetrical accumulation of F-actin are necessary for establishment of cell polarity in the early development of monospores from the marine red alga *Porphyra yezoensis*. *Journal of Experimental Botany*, *59*, 3575-3586.

Li, L., Saga, N. and Mikami, K. (2009b). Ca^{2+} influx and phosphoinositide signalling are essential for the establishment and maintenance of cell polarity in monospores from the red alga *Porphyra yezoensis*. *Journal of Experimental Botany*, *60*, 3477-3489.

Lin, Z. F., Arciga-Reyes, L., Zhong, S. L., Alexander, L., Hackett, R., Wilson, I. and Grierson, D. (2008). SlTPR1, a tomato tetratricopeptide repeat protein, interacts with the ethylene receptors NR and LeETR1, modulating ethylene and auxin responses and development. *Journal of Experimental Botany*, *59*, 4271-4287.

Liu, Q. Y. and Reith, M. (1996a). A sporophyte cell wall protein of the red alga *Porphyra purpurea* (Rhodophyta) is a novel member of the chymotrypsin family of serine proteases. *Journal of Phycology*, *32*, 1003-1009.

Liu, Q. Y., Ross, N., Lanthier, P. and Reith, M. (1996b). A gametophyte cell wall protein of the red alga *Porphyra purpurea* (Rhodophyta) contains four apparent polysaccharide-binding domains. *Journal of Phycology*, *32*, 995-1003.

Loovers, H. M., Postma, M., Keizer-Gunnink, I., Huang, Y. E., Devreotes, P. N., van Haastert, P. J. (2006). Distinct roles of PI(3,4,5)P_3 during chemoattractant signaling in *Dictyostelium*: a quantitative in vivo analysis by inhibition of PI3-kinase. *Molecular Biology of the Cell*, *17*, 1503-1513.

Lumbreras, V., Stevens, D. R. and Purton, S. (1998). Efficient foreign gene expression in *Chlamydomonas reinhardtii* mediated by an endogenous intron. *Plant Journal*, *14*, 441-447.

Manavalan, L. P., Guttikonda, S. K., Tran, L. S. and Nguyen, H. T. (2009). Physiological and molecular approaches to improve drought resistance in soybean. *Plant and Cell Physiology*, *50*, 1260-1276.

Martin, K., Kopperud, K., Chakrabarty, R., Banerjee, R., Brooks, R. and Goodin, M. M. (2009). Transient expression in *Nicotiana benthamiana* fluorescent marker lines provides enhanced definition of protein localization, movement and interactions in planta. *Plant Journal 59*, 150-162.

Masclaux, F., Charpenteau, M., Takahashi, T., Pont-Lezica, R., and Galaud, J. P. (2004). Gene silencing using a heat-inducible RNAi system in *Arabidopsis*. *Biochemical and Biophysical Research Communications*, *321*,364-369.

Matsuyama-Serisawa, K., Yamamoto, M., Fujishita, M., Endo, H., Serisawa, Y., Tabata, S., Kawano, S. and Saga N. (2007). DNA content of the cell nucleus in the macroalga *Porphyra yezoensis* (Rhodophyta). *Fisheries Science*, *73*, 738-740.

Matsuzaki, M., Misumi, O., Shin-I, T., Maruyama, S., Takahara, M., Miyagishima, S. Y., Mori, T., Nishida, K., Yagisawa, F., Yoshida, Y., Nishimura, Y., Nakao, S., Kobayashi, T., Momoyama, Y., Higashiyama, T., Minoda, A., Sano, M., Nomoto, H., Oishi, K., Hayashi, H., Ohta, F., Nishizaka, S., Haga, S., Miura, S., Morishita, T., Kabeya, Y., Terasawa, K., Suzuki, Y., Ishii, Y., Asakawa, S., Takano, H., Ohta, N., Kuroiwa, H., Tanaka, K., Shimizu, N., Sugano, S., Sato, N., Nozaki, H., Ogasawara, N., Kohara, Y. and Kuroiwa, T. (2004). Genome sequence of the ultrasmall unicellular red alga *Cyanidioschyzon merolae* 10D. *Nature*, *428*, 653-657.

McElroy, D., Blowers, A. D., Jenes, B. and Wu, R. (1991). Construction of expression vectors based on the rice actin 1 (*Act1*) 5' region for use in monocot transformation. *Molecular and General Genetics*, *231*, 150-160.

Miao, Y. and Jiang, L. (2007). Transient expression of fluorescent fusion proteins in protoplasts of suspension cultured cells. *Nature Protocols*, *2*, 2348-2353.

Mikami, K., Uji, T., Li, L., Takahashi, M., Yasui, H. and Saga, N. (2009). Visualization of Phosphoinositides via the development of the transient expression system of a cyan fluorescent protein in the red alga *Porphyra yezoensis*. *Marine Biotechnology*, *11*, 563-569.

Miki, B. and McHugh, S. (2004). Selectable marker genes in transgenic plants: applications, alternatives and biosafety. *Journal of Biotechnology*, *107*, 193-232.

Minoda, A., Sakagami, R., Yagisawa, F., Kuroiwa, T. and Tanaka, K. (2004). Improvement of culture conditions and evidence for nuclear transformation by homologous recombination in a red alga, *Cyanidioschyzon merolae* 10D. *Plant and Cell Physiology*, *45*, 667-671.

Mukai, L. S., Craigie, J. S. and Brown, R. G. (1981). Chemical composition and structure of the cell walls of the conchocelis and thallus phases of *Porphyra tenera* (Rhodophyceae). *Journal of Phycology*, *17*, 192-198.

Nakashima, K., Ito, Y. and Yamaguchi-Shinozaki, K. (2009). Transcriptional regulatory networks in response to abiotic stresses in Arabidopsis and grasses. *Plant Physiology*, *149*, 88-95.

Newell, C. A. (2000). Plant transformation technology: Developments and applications. *Molecular Biotechnology*, *16*, 53-65.

Nikaido, I., Asamizu, E., Nakajima, M., Nakamura, Y., Saga, N. and Tabata, S. (2000). Generation of 10,154 expressed sequence tags from a leafy gametophyte of a marine red alga, *Porphyra yezoensis*. *DNA Research, 7*, 223-227.

Nishio, M., Watanabe, K. I., Sasaki, J., Taya, C., Takasuga, S., Iizuka, R., Balla, T., Yamazaki, M., Watanabe, H., Itoh, R., Kuroda, S., Horie, Y., Forster, I., Mak, T. W., Yonekawa, H., Penninger, J. M., Kanaho, Y., Suzuki, A. and Sasaki, T. (2007). Control of cell polarity and motility by the PtdIns(3,4,5)P-3 phosphatase SHIP1. *Nature Cell Biology*, *9*, 36-44.

Niwa, Y., Hirano, T., Yoshimot,o K., Shimizu, M. and Kobayashi, H. (1999). Non-invasive quantitative detection and applications of non-toxic, S65T-type green fluorescent protein in living plants. *Plant Journal*, *18*, 455-463

Nowak, K., Luniak, N., Meyer, S., Schulze, J., Mendel, R. R. and Hansch, R. (2004). Fluorescent proteins in poplar: A useful tool to study promoter function and protein localization. *Plant Biology*, *6*, 65-73.

Ohnuma, M., Misumi, O., Fujiwara, T., Watanabe, S., Tanaka, K. and Kuroiwa, T. (2009). Transient gene suppression in a red alga, *Cyanidioschyzon merolae* 10D. *Protoplasma*, *236*, 107-112.

Ohnuma, M., Yokoyama, T., Inouye, T., Sekine, Y. and Tanaka, K. (2008). Polyethylene glycol (PEG)-mediated transient gene expression in a red alga, *Cyanidioschyzon merolae* 10D. *Plant and Cell Physiology*, *49*, 117-120.

Okano, Y., Aono, N., Hiwatashi, Y., Murata, T., Nishiyama, T., Ishikawa, T., Kubo, M. and Hasebe, M. (2009). A polycomb repressive complex 2 gene regulates apogamy and gives evolutionary insights into early land plant evolution. *Proceedings of the National Academy of Sciences of the United States of America*, *106*, 16321-16326.

Okauchi, M. and Mizukami, Y. (1999). Transient β-Glucuronidase (GUS) gene expression under control of *CaMV 35S* promoter in *Porphyra tenera* (Rhodophyta). *Bulletin of National Research Institute of Aquaculture*, *1*, 13-18.

Perez, P., Tiraby, G., Kallerhoff, J. and Perret, J. (1989). Phleomycin resistance as a dominant selectable marker for plant-cell transformation. *Plant Molecular Biology*, *13*, 365-373.

Piao, T., Luo, M., Wang, L., Guo, Y., Li, D., Li, P., Snell, W. J. and Pan, J. (2009). A microtubule depolymerizing kinesin functions during both flagellar disassembly and flagellar assembly in *Chlamydomonas*. *Proceedings of the National Academy of Sciences of the United States of America*, *106*, 4713-4718.

Radchuk, V. V., Ryschka, U., Schumann, G. and Klocke, E. (2002). Genetic transformation of cauliflower (Brassica oleracea var. botrytis) by direct DNA uptake into mesophyll protoplasts. *Physiologia Plantarum*, *114*, 429-438.

Rao, A. Q., Bakhsh, A., Kiani, S., Shahzad, K., Shahid, A. A., Husnain, T. and Riazuddin, S. (2009). The myth of plant transformation. *Biotechnology Advances*, *27*, 753-763.

Riaño-Pachón, D. M., Corrêa, L. G., Trejos-Espinosa, R. and Mueller-Roeber, B. (2008). Green transcription factors: a chlamydomonas overview. *Genetics*, *179*, 31-39.

Rochaix, J. D. (1995). *Chlamydomonas reinhardtii* as the photosynthetic yeast. *Annual Review of Genetics*, *29*, 209-230.

Saga, N. and Kitade, Y. (2002). *Porphyra*: A model plant in marine sciences. *Fisheries Science*, *68* (Supplement), 1075-1078.

Schroda, M., Blocker, D. and Beck, C. F. (2000). The HSP70A promoter as a tool for the improved expression of transgenes in *Chlamydomonas*. *Plant Journal*, *21*, 121-131.

Serrano, G., Herrera-Palau, R., Romero, J. M., Serrano, A., Coupland, G. and Valverde, F. (2009). Chlamydomonas *CONSTANS* and the evolution of plant photoperiodic signaling. *Current Biology*, *19*, 359-368.

Shimogawara, K., Fujiwara, S., Grossman, A. and Usuda, H. (1998). High-efficiency transformation of *Chlamydomonas reinhardtii* by electroporation. *Genetics*, *148*, 1821-1828.

Singer, S. D. and Ashton, N. W. (2009). MADS about MOSS. *Plant Signaling and Behavior*, *4*, 111-112.

Sizova, I., Fuhrmann, M. and Hegemann, P. (2001). A *Streptomyces rimosus aphVIII* gene coding for a new type phosphotransferase provides stable antibiotic resistance to *Chlamydomonas reinhardtii*. *Gene*, *277*, 221-229.

Stevens, D. R., Rochaix, J. D. and Purton, S. (1996). The bacterial phleomycin resistance gene *ble* as a dominant selectable marker in *Chlamydomonas*. *Molecular and General Genetics*, *251*, 23-30.

Sun, G. H., Zhang, X. C., Sui, Z. H. and Mao, Y. X. (2008). Inhibition of *pds* gene expression via the RNA interference approach in *Dunaliella salina* (Chlorophyta). *Marine Biotechnology*, *10*, 219-226.

Sun, P., Tian, Q. Y., Chen, J. and Zhang, W. H. (2010). Aluminium-induced inhibition of root elongation in *Arabidopsis* is mediated by ethylene and auxin. *Journal of Experimental Botany*, *61*, 347-356.

Szentpetery, Z., Balla, A., Kim, Y. J., Lemmon, M. A., Balla, T. (2009). Live cell imaging with protein domains capable of recognizing phosphatidylinositol 4,5-bisphosphate; a comparative study. *BMC Cell Biology*, *10*, 67.

Takahashi, M., Uji, T., Saga, N. and Mikami, K. (2010). Isolation and regeneration of transiently transformed protoplasts from gametophytic blades of the marine red alga *Porphyra yezoensis*. *Electronic Journal of Biotechnology*, *13*, 2.

Takenaka, M., Yamaoka, S., Hanajiri, T., Shimizu-Ueda, Y., Yamato, K. T., Fukuzawa, H., Ohyama, K. (2000). Direct transformation and plant regeneration of the haploid liverwort *Marchantia polymorpha* L. *Transgenic Research, 9*, 179-185.

Tan, C. P., Qin, S., Zhang, Q., Jiang, P. and Zhao, F. Q. (2005). Establishment of a micro-particle bombardment transformation system for *Dunaliella saliva. Journal of Microbiology, 43*, 361-365.

ten Lohuis, M. R. and Miller, D. J. (1998). Genetic transformation of dinoflagellates (*Amphidinium* and *Symbiodinium*): expression of GUS in microalgae using heterologous promoter constructs. *Plant Journal, 13*, 427-435.

Thonpho, A., Sereeruk, C., Rojvirat, P. and Jitrapakdee, S. (2010). Identification of the cyclic AMP responsive element (CRE) that mediates transcriptional regulation of the pyruvate carboxylase gene in HepG2 cells. *Biochemical and Biophysical Research Communications, 393*, 714-719.

Tian, L. N., Charest, P. J., Seguin, A. and Rutledge, R. G. (2000). Hygromycin resistance is an effective selectable marker for biolistic transformation of *Black spruce* (Picea mariana). *Plant Cell Reports, 19*, 358-362.

Trong-Tue, N., Thao, D. T. P. and Yamaguchi, M. (2010). Role of DREF in transcriptional regulation of the *Drosophila p53* gene. *Oncogene, 29*, 2060-2069.

Uji, T., Takahashi, M., Saga, N. and Mikami, K. (2010). Visualization of nuclear localization of transcription factors with cyan and green fluorescent proteins in the red alga *Porphyra yezoensis*. *Marine Biotechnology, 12*, 150-159.

van Leeuwen, W., Vermeer, J. E. M., Gadella, T. W. J. and Munnik, T. (2007). Visualization of phosphatidylinositol 4,5-bisphosphate in the plasma membrane of suspension-cultured tobacco BY-2 cells and whole *Arabidopsis* seedlings. *Plant Journal, 52*, 1014-1026.

Vermeer, J. E. M., Thole, J. M., Goedhart, J., Nielsen, E., Munnik, T. and Gadella, T. W. (2009). Imaging phosphatidylinositol 4-phosphate dynamics in living plant cells. *Plant Journal, 57*, 356-372.

Waaland, J. R., Stiller, J. W. and Cheney, D. P. (2004). Macroalgal candidates for genomics. *Journal of Phycology, 40*, 26-33.

Walker, T. L., Collet, C. and Purton, S. (2005). Algal transgenics in the genomic ERA. *Journal of Phycology, 41*, 1077-1093.

Williams, M. E., Torabinejad, J., Cohick, E., Parker, K., Drake, E. J., Thompson, J. E., Hortter, M. and DeWald, D. B. (2005). Mutations in the arabidopsis phosphoinositide phosphatase gene SAC9 lead to over accumulation of PtdIns(4,5)P-2 and constitutive expression of the stress-response pathway. *Plant Physiology, 138*, 686-700.

Xu, Z. S., Ni, Z. Y., Liu, L., Nie, L. N., Li, L. C., Chen, M. and Ma, Y. Z. (2008). Characterization of the TaAIDFa gene encoding a CRT/DRE-binding factor responsive to drought, high-salt, and cold stress in wheat. *Molecular Genetics and Genomics, 280*, 497-508.

Xue, H. W., Chen, X. and Me, Y. (2009). Function and regulation of phospholipid signallingin plants. *Biochemical Journal, 421*, 145-156.

Yoshida, Y., Kuroiwa, H., Hirooka, S., Fujiwara, T., Ohnuma, M., Yoshida, M., Misumi, O., Kawano, S. and Kuroiwa, T. (2009). The bacterial ZapA-like protein ZED is required for mitochondrial division. *Current Biology, 19*, 1491-1497.

Yoshihara, C., Inoue, K., Schichnes, D., Ruzin, S., Inwood, W. and Kustu, S. (2008). An Rh1-GFP fusion proteins in the cytoplasmic membrane of a white mutant strain of *Chlamydomonas reinhardtii*. *Molecular Plant, 1*, 1007-1020.

Yu, C., Wang, H. S., Yang, S., Tang, X. F., Duan, M. and Meng, Q. W. (2009). Over expression of endoplasmic reticulum omega-3 fatty acid desaturase gene improves chilling tolerance in tomato. *Plant Physiology and Biochemistry, 47*, 1102-1112.

Zalewski, W., Galuszka, P., Gasparis, S., Orczyk, W. and Nadolska-Orczyk, A. (2010). Silencing of the *HvCKX1* gene decreases the cytokinin oxidase/dehydrogenase level in barley and leads to higher plant productivity. *Journal of Experimental Botany, 61*, 1839-1851.

Zaslavskaia, L. A., Lippmeier, J. C., Kroth, P. G., Grossman, A. R. and Apt, K. E. (2000). Transformation of the diatom *Phaeodactylum tricornutum* (Bacillariophyceae) with a variety of selectable marker and reporter genes. *Journal of Phycology, 36*, 379-386.

Zhou, P., Yang, F., Yu, J. J., Ao, G. M. and Zhao, Q. (2010). Several *cis*-elements including a palindrome involved in pollen-specific activity of *SBgLR* promoter. *Plant Cell Reports, 29*, 503-511.

Zhu, Y., Li, Z. Y., Xu, B., Li, H. D., Wang, L. J., Dong, A. W. and Huang, H. (2008). Subcellular localizations of AS1 and AS2 suggest their common and distinct roles in plant development. *Journal of Integrative Plant Biology, 50*, 897-905.

In: Porphyra yezoensis
Editor: Koji Mikami

ISBN 978-1-61122-304-0

Chapter 7

HEAVY-ION BEAM MUTAGENESIS IN *PORPHYRA YEZOENSIS*

Kyosuke Niwa[1*] and Tomoko Abe[2]

[1]Fisheries Technology Institute, Hyogo Prefectural Technology Center for Agriculture, Forestry and Fisheires, Akashi, Hyogo, 674-0093, Japan

[2]RIKEN Nishina Center, Wako, Saitama, 351-0198, Japan

ABSTRACT

Cross-fertilization experiment is difficult for the marine crop *Porphyra yezoensis* because of its monoecious nature and because of the few morphological markers to confirm out-crossing. Thus, the pigmentation mutants of *Porphyra yezoensis*, the colors of which are useful as genetic markers, are attractive experimental materials for genetic research. In this chapter, we describe the principal of heavy-ion beam mutagenesis and the isolation method of artificial mutants in *P. yezoensis*. Furthermore, we discuss the contribution of the various pigmentation mutants and other mutants (such as morphological and physiological mutants) to basic biological and applied breeding studies.

INTRODUCTION

The marine crop *Porphyra yezoensis*, commonly known as nori, is mainly cultivated in Japan, Korea and China. The total annual production value of nori in Japan amounts to almost US$1 billion. Thus, *P. yezoensis* is one of the most lucrative marine crops worldwide, and nori cultivators have always request new breeds of cultivars with higher yield, better flavor of dried nori, higher tolerance to disease and high-temperature, and deeper color even under low-nutrient conditions in nori farms. For further development of new cultivars, it is necessary to estimate genetic knowledge on the marine crop. However, since cross-fertilization is difficult for *P. yezoensis* because of its monoecious nature and few

* Author for correspondence: e-mail kyousuke_niwa@pref.hyogo.lg.jp.

morphological markers to confirm out-crossing, there was no genetic study in this species until finding the spontaneous pigmentation mutants.

In the 1970s, Kobara et al. (1976) isolated the spontaneous green mutant of *P. yezoensis* for the first time from a sectored blade (chimera) in laboratory culture during the course of their studies on the red mutant strain established by nori farmers through selection from cultivated populations. Because the color of the pigmentation mutants can be used as a genetic marker, genetic research of *Porphyra* progressed by cross-fertilization experiments among the wild type and some spontaneous pigmentation mutants (Miura and Kunifuji, 1980; Miura, 1985; Ohme et al., 1986; Niwa et al., 1993, 2002). From the results, inheritance patterns of their color were in accordance with the Mendel's laws, and conchospores that were released from conchosporangia in heterozygous conchocelis developed into sectored F_1 gametophytic blades (chimeras) in high frequency consisting of both parental colors in two to four sectors (Figure 1). On the basis of the phenomenon, it has been revealed that the members of *Porphyra* undergo meiosis during the first two divisions of the germinating conchospores (Ma and Miura, 1984; Burzycki and Waaland, 1987; Ohme and Miura, 1988; Mitman and van der Meer, 1994; Yan et al., 2005a), and it can be considered that the germinating conchospores correspond to the ordered tetrad in *Neurospora* (Ohme and Miura, 1988). Therefore, gene-centromere distances and linkage relationships between mutant genes have been determined by tetrad analyses in *P. yezoensis* (Ohme and Miura, 1988; Niwa et al., 1993, 2002). Furthermore, based on the genetic knowledge, a new cultivar was established by crossbreeding using two pigmentation mutants of *P. yezoensis* (Miura and Shin, 1989). Therefore, for further development of genetic and breeding studies of cultivated *Porphyra*, it is more important to induce and isolate the various mutants.

The mutants of *P. yezoensis* are attractive experimental materials not only for genetic and breeding studies but also for physiological and biochemical studies. In particular, blade color of *Porphyra* is one of the very important factors in determining the quality of commercial dried nori sheets: a deep black color is considered to be high grade in Japan. Thus, the pigmentation mutants of *P. yezoensis* have been used in physiological and biochemical studies (Kikuchi et al., 1979; Merrill et al., 1983; Aruga and Miura, 1984; Kato and Aruga, 1984; Abe et al., 1994).

After first trial of chemical mutagenesis in *P. yezoensis* (Katayama, 1983, 1984), artificial pigmentation mutants of *Porphyra* have been induced by chemical mutagens (Mitman and van der Meer, 1994; Yan and Aruga, 1997, 2000; Yan et al., 2000; Wang et al., 2008) and gamma rays (Wang et al., 2000, Yan et al., 2005b). Recently, we demonstrated that heavy-ion beam irradiation is an effective tool for isolation of artificial mutants of *P. yezoensis* (Niwa et al., 2009a). In this chapter, we describe the principal of heavy-ion beam mutagenesis, isolation method and characterization of the artificial mutants in *P. yezoensis*.

PRINCIPAL OF HEAVY-ION BEAM MUTAGENESIS

In mutagenesis, genes are artificially damaged using chemical mutagens, gamma rays, X-rays, and a variety with the desired characteristics is selected from among the resulting mutants. Mutagenesis using radiation is based on the same principal as naturally and commonly occurring mutations from cosmic rays that impinge on the Earth from outer space.

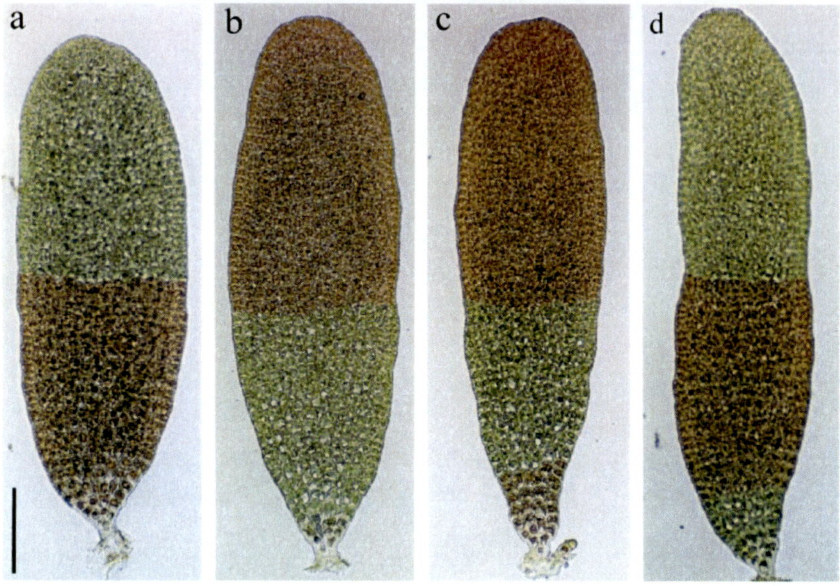

Figure 1. Two- and three-sectored F_1 gametophytic blades (chimeras) from heterozygous conchocelis in the cross between the green mutant and the wild type of *Porphyra yezoensis*. Scale bar 0.1 mm in a also applies to b-d. (Quoted from Niwa et al. [6]).

Ion beams are accelerated ions (charged atoms) with atomic nuclei. Ions are different from photons, gamma rays and X-rays, since they have a mass and electric charge. By contrast photons, gamma and X-rays have neither. As a result, ion beams provide a different linear energy transfer (LET). The choice of ion beam depends on the characteristics of the ion with respect to electrical charge and velocity. When a high LET is required, a heavier, highly charged ion is selected with a low velocity. For example, this is 0.2 keV/μm for gamma rays and 2 keV/μm for X-rays, but is much larger and widely varying for ion beams, e.g., 23 to 285 keV/μm for $^{12}C^{6+}$ ion and 640 to 4000 keV/μm for $^{56}Fe^{24+}$ ion. High-LET radiation, such as ion beams, can produce dense and localized ionization in matters along their tracks, in contrast to low-LET radiation (gamma rays and X-rays). Ion beam can produce double-stranded DNA breaks in cells, which are not easily repaired and result in mutation more effectively than low-LET radiation. The biological effects of high-LET radiation are generally greater than those of low-LET radiation.

Ion beams usually are generated through particle accelerators. The RIKEN RI-Beam Factory (RIBF, Figure 2) is a multi-disciplinary facility for nuclear physics. The light ion beam, up to 135 MeV/nucleon (about half the speed of light) was established since 1986. Shortly after it was used in cancer therapy, collaboration was established among radiation oncologists, physicists, and biologists. Plant scientists started using the RIBF for radiation biology research from 1989. Plant breeding trials began in 1993. At present, biological samples are irradiated with ^{12}C, ^{14}N, ^{20}Ne, ^{40}Ar and ^{56}Fe beams. The beam accelerated by the RIKEN Ring Cyclotrons is transported to the E5B beam line, which is used exclusively for radiation of biology experiments. Ion beam irradiates samples via an automatic irradiation system, which consists of a range sifter and an automatic sample changer. The LET of the beam incident on the samples can be adjusted using the range shifter (Figure 2, Ryuto et al., 2006).

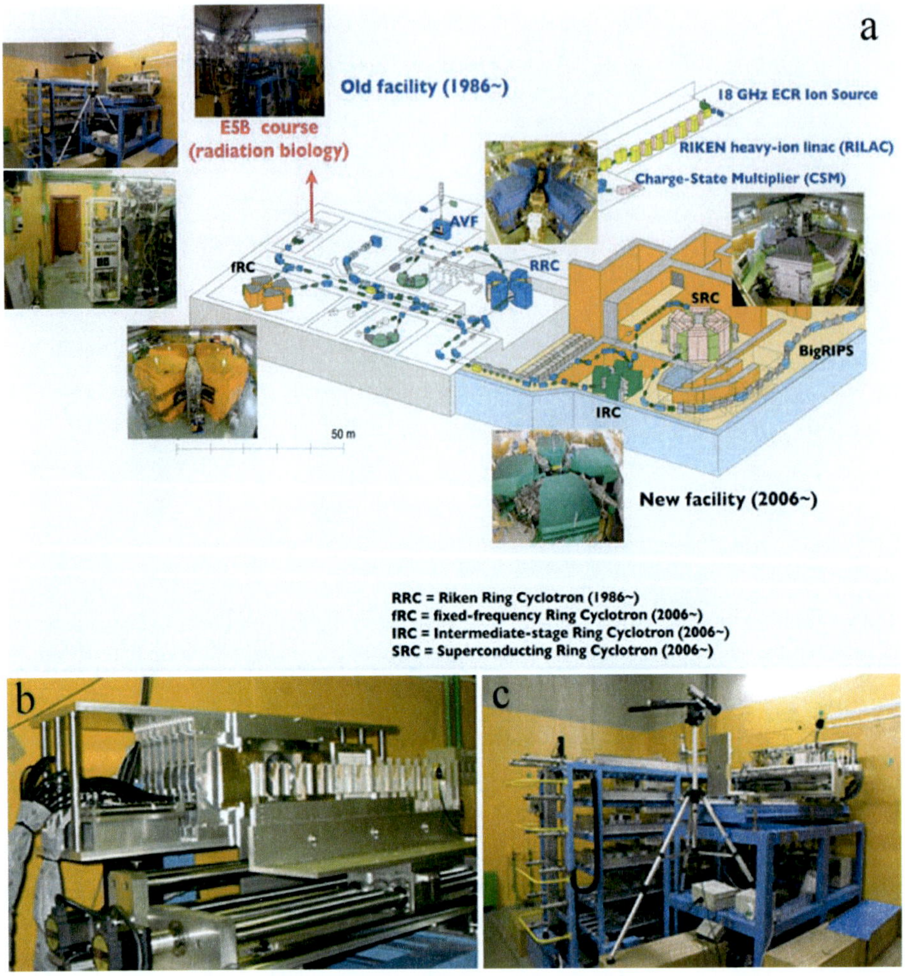

Figure 2. A schematic view of the RIKEN RI-Beam Factory (RIBF, a) and photographs of a range sifter (b) and an automatic sample changer (c) in the automatic irradiation system of heavy-ion beams.

Initial studies showed that ion beams are highly effective in inducing mutations using developing seed embryos at a particular stage after fertilization without damage to other plant tissues (Abe et al., 1995). Many types of mutants including albino (Bae et al., 2001), variegation, herbicide- and salt-tolerant phenotypes in M_1 plants in tobacco (Abe et al., 2000). Exposure of plant materials to an ion beam for a few seconds to minutes triggered various genetic changes. Advantages of ion beam mutagenesis include: low dose with high survival rates, induction of high mutation rates and a wide range of variation (Table 1). The use ion beam also causes less damage to the plants and breaks fewer genes, thus requiring less time for mutations to stabilize. The time span for breeding can be shortened significantly to two to three years. The approach has been particularly successful in flower breeding. For instance, RIKEN have put 16 new flower cultivars in the market in Japan, USA, Canada and EU since 2001. The ion beam is, therefore, an excellent tool also for mutagenesis of *P. yezoensis* with high efficiency.

Table 1. Mutant lines developed in various crops using RIBF

Mutant phenotype/plants (generation)	Variety, strain/Plant material treated	Survival rate (%)[1]	Mutation rate (%)
Sterile			
Verbena	Temari Coral Pink/stem node containing lateral buds	84-90	0.9-2.8
Eucalyptus	Shoot primordia	80	9.3
Chlorophyll deficient			
Tobacco (M_1)	BY-4/ovary containing fertilization embryos	75	0.6
Rice	Nipponbare/imbibed seeds	65-86	6.2-10.3
Flower colour and shape			
Dahlia	Miharu/shoots	NE	20.3
Rose	Bridal Fantasy/dormant scions	NE	51.7
Chrysanthemum	Chiyo/stem	93.7	14
Dwarf			
Barley	Kantohadaka 77/dry seeds	ND	0.08
Pepper (M_1)	California Wonder/dry seeds	80	1.3
Millet	Mojappe/dry seeds	ND	0.1
Buckwheat	Shinano natsu soba/dry seeds	NE	0.6
	Rotundatiem/dry seeds	NE	1.1

[1] ND: no data, NE: no effect.

Isolation Method of Artificial Pigmentation Mutants

P. yezoensis has a heteromorphic alternation of generation with gametophytic haploid blades and sporophytic diploid filaments called conchocelis (Figure 3); thus, haploid blades seem to be suitable for inducing mutation in *Porphyra* because recessive mutations appear directly in the phenotype. In addition, *P. yezoensis* propagates by an asexual subcycle through archeospores (monospores) released from gametophytic blades. Therefore, if an archeospore that was released from a mutated cell cluster developed into a mature blade, a pure line (complete homozygous conchocelis strain) with the mutational genotype is easily established by self-fertilization.

To induce mutation, the gametophytic blades of *P. yezoensis*, which were put into centrifuge tubes (15 mL) containing seawater, were irradiated by $^{12}C^{+6}$ beams (135 MeV/nucleon, about 50% of speed of light) in the automatic irradiation system (Figure 2). The LET of ions corresponded to 23 keV/μm. When the clonal blades from the archeospores after approximately 1.5 months of culture were irradiated with the $^{12}C^{+6}$ beams at doses of 25, 50, 100, 150, 200, 300 and 400 Gy, the survival rate of gametophytic cells one day after heavy-ion beam irradiation was more than 97% even at 400 Gy (Niwa et al., 2009a). However, the survival rate two weeks after irradiation decreased at 200, 300 and 400 Gy (Niwa et al., 2009a). Furthermore, the blades irradiated at 150 Gy or less were normally growing, but the blades irradiated at 200 Gy or more didn't grow well (Figure 4, Niwa et al., 2009a). These results suggest that a dose of 150 Gy or less is suitable to induce mutation in LET of 23 keV/μm with $^{12}C^{+6}$ beams for the isolation of mutants of *P. yezoensis* (Niwa et al., 2009a).

Figure 3. Gametophytic haploid blades (a) and diploid conchocelis (b) of *Porphyra yezoensis*.

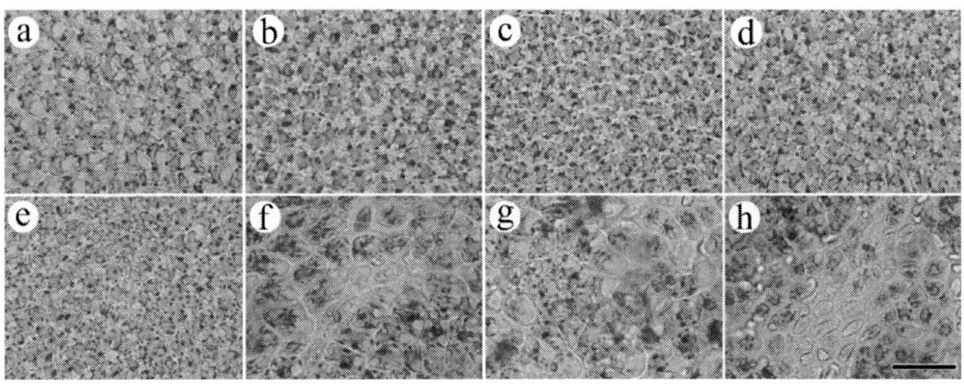

Figure 4. Gametophytic cells of *Porphyra yezoensis* 20 days after heavy-ion beam irradiation. (a) Control (no irradiation). (b) Irradiation at 25 Gy. (c) Irradiation at 50 Gy. (d) Irradiation at 100 Gy. (e) Irradiation at 150 Gy. (f) Irradiation at 200 Gy. (g) Irradiation at 300 Gy. (h) Irradiation at 400 Gy. Scale bar 50 μm in h also applies to a-g. (Quoted from Niwa et al. [26]).

Since it was difficult to discriminate objectively between mutational cells and non-mutational cells in the irradiated blades, the mutational rate after heavy-ion beam irradiation has been not investigated in *P. yezoensis*. However, mutations induced by ion-beam irradiation at the molecular level have been most extensively studied in mammalian cells. The frequency of deletion is higher for ion beam irradiation than for gamma rays (e.g., Suzuki et al., 1996). In the case of *Arabidopsis* irradiated with $^{12}C^{6+}$ ions, half of the mutants produced had point-like mutations and the other half large DNA alterations such as inversions, insertions, translocations and large deletions (Shikazono et al., 2005). An LET of 30 keV/μm was found to be the most effective for mutation induction in *Arabidopsis* (Kazama et al., 2008). In the case of rice, the highest mutation was observed in LET from 61 to 74 keV/μm with $^{12}C^{6+}$ and $^{20}Ne^{10+}$ beams (Hayashi et al., 2008). Thus, the LET of ion beams seems to be an important factor affecting mutagenesis. To induce various mutants in *P. yezoensis* with high efficiency, mutational rates of the gametophytic cells irradiated at different LET condition also need to be examined based on the objective criteria in the future.

Isolation method of the artificial pigmentation mutant in *P. yezoensis* after irradiation is shown in Figure 5. The procedure after other mutagenetic treatments, such as chemical mutagenesis and gamma rays, is also basically the same. Small cell clusters with mutation colors were found in several irradiated blades under a light microscope about 20 days after irradiation (Figure 6). The blade piece containing a mutated cell cluster was cut out from the irradiated blade. The blade piece was cultured in a flask, and vinylon monofilaments were placed in the flask to attach archeospores that were released from the mutated cell cluster. The archeospore germling grew to a single-colored blade that was a clonal blade of the mutated cell (Figure 7, Niwa et al., 2009a). The single blade was cultured in a flask until maturation. After self-fertilization, the mutant strain was established as a conchocelis strain. Using above-mentioned method, we have succeeded in isolation of genetically stable pigmentation mutants of *P. yezoensis* (Niwa et al., 2009a). In this procedure, dozens of cell clusters with mutation colors were observed under the light microscope after gametophytic blades were irradiated by heavy-ion beams; however, most of the cell clusters did not release archeospores. If archeospores are always released from all mutated cell clusters, more artificial mutants of *P. yezoensis* will be easily isolated. In recent years, Mizuta et al. (2003) developed a simple method using allantoin to produce archeospores from gametophytic blades of *P. yezoensis*; thus, the method will be useful to isolate the artificial mutants from the mutated cell clusters more efficiently.

Although we demonstrated the isolation method of artificial pigmentation mutants from gametophytic blades of *P. yezoensis*, it is possible to induce the mutant from conchocelis. Conchocelis phase of *Porphyra* is diploid microscopic sporophyte, and *P. yezoensis* undergoes meiosis during the first two divisions of the germinating conchospores. Therefore, if mature conchocelis colonies with conchosporangia are irradiated by heavy-ion beams, it is predicted that conchospores from the irradiated mature conchocelis will develop into chimeric blades in high frequency compared with the blades developing from conchospores that are released from normal conchocelis strains without heavy-ion beam irradiation. Using archeospores released from mutated sectors in the chimeric blades, artificial pigmentation mutants will be established from the irradiated conchocelis. In particular, we can maintain conchocelis strains in small bottles for a long period and obtain a lot of conchospores from a small colony of mature conchocelis. Therefore, if mature conchocelis colonies are irradiated by heavy-ion beams and maintained in the laboratory, it will be possible to start laboratory culture anytime for isolation of the mutants at convenient time.

Characterization of Pigmentation Mutants

The gametophytic blades of *P. yezoensis* usually contain chlorophyll *a* (Chl. *a*), phycoerythrin (PE) and phycocyanin (PC) as the major photosynthetic pigments. The blade color is based on the relative amounts of these pigments. Until now, some spontaneous pigmentation mutants (red, green, violet, orange, yellow) of *P. yezoensis* were characterized by *in vivo* absorption spectra and the pigment contents (Aruga and Miura, 1984; Niwa et al., 1993). From the results, Aruga and Miura (1984) inferred that the red mutant is mainly due to qualitative variation of phycobilins, while the green mutant is mainly due to quantitative variation by the low content of PE.

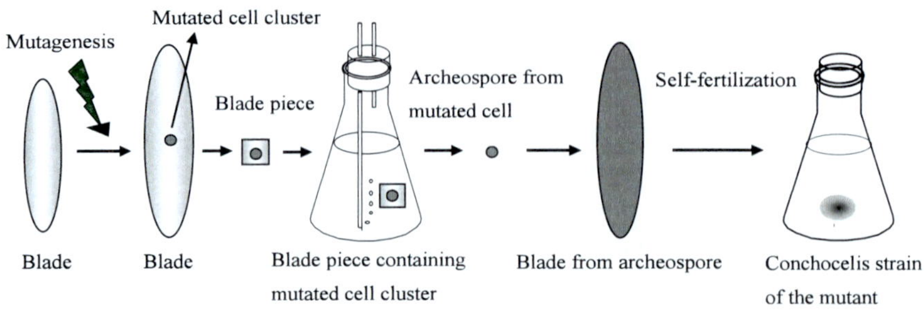

Figure 5. Isolation method of the artificial pigmentation mutant of *Porphyra yezoensis* after mutagenetic treatment.

Figure 6. Small cell clusters having mutation colors in gametophytic blades of *Porphyra yezoensis* about 20 days after heavy-ion beam irradiation. (a) Cluster of red cells. (b) Cluster of green cells. (c) Cluster of deep reddish brown-colored cells. Scale bar 50 μm in c also applies to a and b. (Quoted from Niwa et al. [26]).

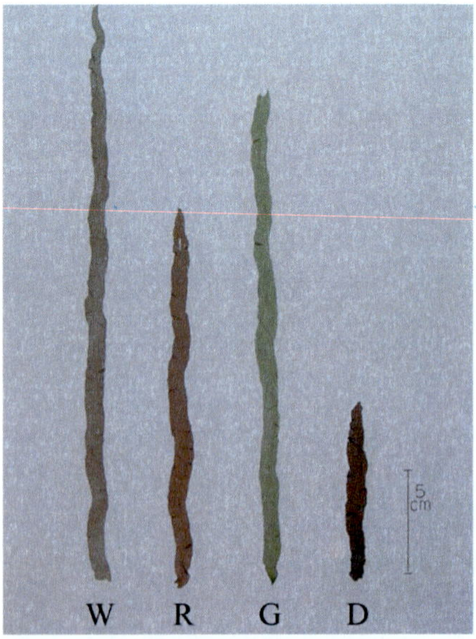

Figure 7. Four color types of gametophytic blades from archeospores released from blades with normal (wild type) or mutated cell clusters in *Porphyra yezoensis* after heavy-ion beam irradiation. W, wild type; R, red mutant; G, green mutant; D, deep reddish brown colored mutant. (Quoted from Niwa et al. [26]).

Niwa et al. (1993) inferred that the violet mutant is mainly due to quantitative variation by the low content of Chl. *a,* while the orange mutant is mainly due to quantitative variation by the low content of PC. Furthermore, Niwa et al. (1993) presumed that the yellow mutant, which is the recombinant with double recessive genes by the cross between the red and green mutant (Miura and Kunifuji 1980, Ohme et al., 1986), is due to qualitative and quantitative variations of phycobilins, indicating that the mutant has both characteristics of the red and green mutants. Because the blade color was changed by the relative ratios of the three photosynthetic pigments as mentioned above, it is possible to artificially induce various pigmentation mutants in *P. yezoensis* (Yan and Aruga, 1997, 2000; Yan et al., 2000) as well as other red algae (van der Meer, 1990).

Heavy-ion beam mutagenesis also has isolated several pigmentation mutants of *P. yezoensis* (Niwa et al., 2009a; Niwa, unpublished data). Of the mutants, the artificial green mutant showed lower PE content compared with the wild type as reported by previous studies (Figure 8, Niwa et al., 2009a). On the other hand, the artificial red mutant showed higher PE content (Figure 8, Niwa et al., 2009a), and this characteristic is different from previous red mutants. Furthermore, genetic analyses of these artificial pigmentation mutants are carried out by cross-fertilization experiments. These genetic analyses suggest that both of the two mutants are controlled by a single recessive nuclear gene and that they are useful as genetic markers for crossbreeding of *P. yezoensis* (Niwa, 2010).

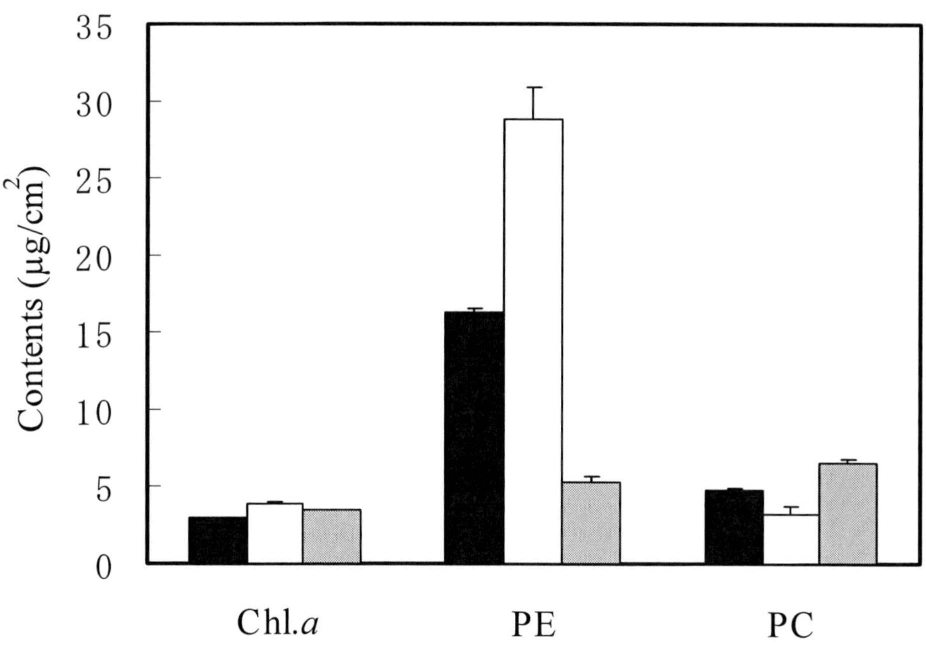

Figure 8. Contents of chlorophyll a (Chl. a), phycoerythrin (PE) and phycocyanin (PC) in the wild type (■), artificial red mutant (□) and artificial green mutant (▒) of *Porphyra yezoensis*. (Quoted from Niwa et al. [26]).

PROSPECT

By heavy-ion mutagenesis, the deep reddish brown-colored blades also developed from the archeospores released from deep reddish brown-colored cell clusters (see Figure 7). Unfortunately, because the deep reddish colored blades did not mature, no conchocelis strain of the mutant, which can be maintained as a pure line, was isolated. However, if deep reddish brown-colored blades with high growth rate and/or other prominent mutants for nori cultivation are induced by heavy-ion beam irradiation, they may be directly used as cultivars in nori farms. Therefore, heavy-ion beam mutagenesis has potential for mutation breeding of this marine crop.

In Japan, *P. yezoensis* has become the main cultivated species (Miura, 1984; Niwa et al., 2004). Although there are many cultivated strains that have been isolated by selective breeding, intensive breeding has led to genetic uniformity in Japanese nori farms (Niwa and Aruga, 2003, 2006; Niwa et al., 2004). Therefore, to establish new cultivars, it is important to collect genetically different strains and to carry out crossbreeding in the marine crop (Niwa et al., 2005a, 2005b, 2008a, 2008b, 2009b; Niwa and Kobiyama, 2009). Furthermore, because our recent study has demonstrated that allopolyploidy occurred in the progeny of interspecific hybridization between *P. yezoensis* and the closely related species *P. tenera*, our findings provide potential for development of polyploidy breeding in the marine crop (Niwa and Sakamoto, 2010; Niwa et al., 2010). Thus, induction and isolation of the pigmentation mutants from various genetically different strains, including *P. yezoensis* and the closely related species, will be more necessary to progress crossbreeding and polyploidy breeding in the marine crop, although molecular markers for confirmation of cross-fertilization recently have been developed in *Porphyra yezoensis* (Sun et al., 2006; Park et al., 2007; Niwa et al., 2010).

Pigmentation mutants of *Porphyra* have been used not only in genetic and breeding studies but also in physiological studies (Kikuchi et al., 1979; Merrill et al., 1983; Aruga and Miura, 1984; Kato and Aruga, 1984; Abe et al., 1994), as well as in biological studies of their life history, such as meiosis and sex determination (Ohme and Miura, 1988; Mitman and van der Meer, 1994; Yan et al., 2005a). In *P. yezoensis*, the haploid gametophytic blade possesses only three chromosomes per single cell (Kito, 1978; Ohme and Miura, 1988), and complete nucleotide sequences of plastid DNA have been determined (Accession number of DNA data, Bank of Japan, AP006715). Moreover, expressed sequence tag (EST) analyses have also been conducted in *P. yezoensis* (Nikaido et al., 2000; Asamizu et al., 2003), and whole genome projects of this species and other *Porphyra* species are progressing in Japan and USA. Accordingly, the genome investigation of *P. yezoensis* has received more attention than any other macroalgae (Waaland et al., 2004). Based on the molecular background, pigmentation mutants and other mutants (such as morphological and physiological mutants) of *P. yezoensis* may be attractive experimental materials to identify molecular genetic mechanism in the future.

CONCLUSION

In this chapter, we demonstrated the practical method for induction and isolation of the artificial pigmentation mutants in *P. yezoensis* by heavy-ion beam irradiation. This mutagenesis will be a power method for isolation of various pigmentation and other mutants that will contribute to basic biological and breeding studies in *P. yezoensis*.

REFERENCES

Abe, S., Murakami, A., Ohki, K., Aruga, Y., and Fujita, Y. (1994). Changes in stoichiometry among PSI, PSII and Cyt b_6-*f* complexes in response to chromatic light for cell growth observed with the red alga *Porphyra yezoensis*. *Plant Cell Physiol.* 35: 901-906.

Abe, T., Yoshida, S., Sakamoto, T., Kameya, T., Kitayama, S., Inabe, N., Kase, M., Goto, A., and Yano, Y. (1995). Is irradiation of heavy ion beams at specific stages of the fertilization cycle of plants effective for mutagenesis? *Proc. of the US-Japanese Joint Meeting*, pp. 469-477.

Abe, T., Bae, C. H., Ozaki, T., Wang, K., and Yoshida, S. (2000). Stress-tolerant mutants induced by heavy-ion beams. *Gamma Field Symp.* 39: 45-54.

Aruga, Y., and Miura, A. (1984). *In vivo* absorption spectra and pigment contents of the two types of color mutants of *Porphyra*. *Jap. J. Phycol.* 32: 243-250.

Asamizu, E., Nakajima, M., Kitade, Y., Saga, N., Nakamura, Y., and Tabata, S. (2003). Comparison of RNA expression profiles between the two generations of *Porphyra yezoensis* (Rhodophyta), based on expressed sequence tag frequency analysis. *J. Phycol.* 39: 923-930.

Bae, C. H., Abe, T., Matsuyama, T., Fukunishi, N., Nagata, N., Nakano, T., Kaneko, Y., Miyoshi, K., Matsushima, H., and Yoshida, S. (2001). Regulation of chloroplast gene expression is affected in *ali*, a novel tobacco albino mutant. *Ann. Bot.* 88: 545-553.

Burzycki, G. M., and Waaland, J. R. (1987). On the position of meiosis in the life history of *Porphyra torta* (Rhodophyta). *Bot. Mar.* 30: 5-10.

Hayashi, Y., Takehisa, H., Kazama, Y., Ichida, H., Ryuto, H., Fukunishi, N., and Abe, T. (2008). Effects of ion beam irradiation on mutation induction in rice. *Cyclotoron and their applications 2007,* pp. 237-239

Katayama, K. (1983). Studies of mutations of *Porphyra*–I. Employment of some chemical mutagens. *Bull. Fish. Res. Stn. Okayama Pref.* 57: 516 (in Japanese).

Katayama, K. (1984). Studies of mutations of *Porphyra*–II. Mutations induced by several chemical mutagens. *Bull. Fish. Res. Stn. Okayama Pref.* 58: 43-49 (in Japanese).

Kato, M., and Aruga, Y. (1984). Comparative studies on the growth and photosynthesis of the pigmentation mutants of *Porphyra yezoensis* in laboratory culture. *Jpn. J. Phycol.* 32: 334-347.

Kazama, Y., Saito, H., Yamamoto, Y. Y., Hayashi, Y., Ichida, H., Ryuto, H., Fukunishi, N., and Abe, T. (2008). LET-dependent effects of heavy-ion beam irradiation in *Arabidopsis thaliana*. *Plant Biotech.* 25: 113-117.

Kikuchi, R., Ashida, K., and Hirao, S. (1979). Phycobilins in different color types of *Porphyra yezoensis* Ueda. *Bull. Jpn. Soc. Sci. Fish.* 45: 1461-1466.

Kito, H. (1978). Cytological studies on genus *Porphyra*. *Bull. Tohoku Res. Fish. Res. Lab.* 39: 29-84 (in Japanese with English abstract).

Kobara, T., Miura, A., and Aruga, Y. (1976). *In vitro* studies on the green type mutant of *Porphyra yezoensis* Ueda. *La mer* 14: 58-63 (in Japanese with English abstract).

Ma, J. H., and Miura, A. (1984). Observations of the nuclear division in the conchospores and their germlings in *Porphyra yezoensis* Ueda. *Jpn. J. Phycol.* 32: 373-378 (in Japanese with English abstract).

Merrill, J. E., Mimuro, M., Aruga, Y., and Fujita, Y. (1983). Light-harvesting for photosynthesis in four strains of the red alga *Porphyra yezoensis* having different phycobilins contents. *Plant Cell Physiol.* 24: 261-266.

Mitman, G. G., and van der Meer, J. P. (1994). Meiosis, blade development, and sex determination in *Porphyra purpurea* (Rhodophyta). *J. Phycol.* 30: 147-159.

Miura, A. (1984). A new variety and a new form of *Porphyra* (Bangiales, Rhodphyta) from Japan: *Porphyra tenera* Kjellman var. *tamatsuensis* Miura, var. nov. and *P. yezoensis* Ueda form. *narawaensis* Miura, form. nov. *J. Tokyo Univ. Fish.* 71: 1-14, pl. I-XI.

Miura, A. (1985). Genetic analysis of the variant color types of light red, light green and light yellow phenotypes of *Porphyra yezoensis* (Rhodophyta, Bangiaceae). In H. Hara (Ed.), *Origin and Evolution of Diversity in Plants and Plant Communities* (edition, pp. 270-284). Tokyo, Japan: Academia Scientific Book.

Miura, A., and Kunifuji, Y. (1980). Genetic analysis of the pigmentation variants in *Porphyra yezoensis*. *Iden* 1980: 34 (9): 14–20 (in Japanese).

Miura, A., and Shin, J.-A. (1989). Crossbreeding in cultivars of *Porphyra yezoensis* (Bangiales, Rhodophyta)–Preliminary report. *Korean J. Phycol.* 4: 207-211.

Mizuta, H., Yasui, H., and Saga, N. (2003). A simple method to mass produce monospores in the thallus of *Porphyra yezoensis* Ueda. *J. Appl. Phycol.* 15: 345-349.

Nikaido, I., Asamizu, E., Nakajima, M., Nakamura, Y., Saga, N., and Tabata, S. (2000). Generation of 10,154 expressed sequence tags from a leafy gametophyte of a marine red alga, *Porphyra yezoensis*. *DNA Res.* 7: 223-227.

Niwa, K., Miura, A., Shin, J-A., and Aruga, Y. (1993). Characterization and genetic analysis of the violet type pigmentation mutant of *Porphyra yezoensis* Ueda (Bangiales, Rhodophyta). *Korean J. Phycol.* 8: 217–230.

Niwa, K., Mizuta, A., and Aruga, Y. (2002). Genetic characterization of a spontaneous green-type pigmentation mutant of *Porphyra yezoensis* and the significance of using heterozygous conchocelis in nori farming. *Fish. Sci.* 68: 729-735.

Niwa, K., and Aruga, Y. (2003). Rapid DNA extraction from conchocelis and ITS-1 rDNA sequences of seven strains of cultivated *Porphyra yezoensis* (Bangiales, Rhodophyta). *J. Appl. Phycol.* 15: 29-35.

Niwa, K., Kikuchi, N., Iwabuchi, M., and Aruga, Y. (2004). Morphological and AFLP variation of *Porphyra yezoensis* Ueda form. *narawaensis* Miura (Bangiales, Rhodophyta). *Phycol. Res.* 52: 180-190.

Niwa, K., Kikuchi, N., and Aruga, Y. (2005a). Morphological and molecular analysis of the endangered species *Porphyra tenera* (Bangiales, Rhodophyta). *J. Phycol.* 41: 294-304.

Niwa, K., Kobiyama, A., and Aruga, Y. (2005b). Confirmation of cultivated *Porphyra tenera* (Bangiales, Rhodophyta) by polymerase chain reaction restriction fragment length polymorphism analyses of the plastid and nuclear DNA. *Phycol. Res.* 53: 296-302.

Niwa, K., and Aruga, Y. (2006). Identification of currently cultivated *Porphyra* species by PCR-RFLP analysis. *Fish. Sci.* 72: 143-148.

Niwa, K., Furuita, H., Yamamoto, T., and Kobiyama, A. (2008a). Identification and characterization of a green-type mutant of *Porphyra tenera* Kjellman var. *tamatsuensis* Miura (Bangiales, Rhodophyta). *Aquaculture* 274: 126-131.

Niwa, K., Kato, A., Kobiyama, A., Kawai, H., and Aruga, Y. (2008b). Comparative study of wild and cultivated *Porphyra yezoensis* (Bangiales, Rhodophyta) based on molecular and morphological data. *J. Appl. Phycol.* 20: 261-270.

Niwa, K., Hayashi, Y., Abe, T., and Aruga, Y. (2009a). Induction and isolation of pigmentation mutants of *Porphyra yezoensis* (Bangiales, Rhodophyta) by heavy-ion beam irradiation. *Phycol. Res.* 57: 194-202.

Niwa, K., Iida, S., Kato, A., Kawai, H., Kikuchi, N., Kobiyama, A., and Aruga, Y. (2009b). Genetic diversity and introgression in two cultivated species (*Porphyra yezoensis* and *Porphyra tenera*) and closely related wild species of *Porphyra* (Bangiales, Rhodophyta). *J. Phycol.* 45: 493-502.

Niwa, K., and Kobiyama, A. (2009). Simple molecular discrimination of cultivated *Porphyra* species and related wild species (Bangiales, Rhodophyta). *Phycol. Res.* 57: 299–303.

Niwa, K. (2010). Genetic analysis of artificial green and red mutants of *Porphyra yezoensis* Ueda (Bangiales, Rhodophyta). *Aquaculture* 308: 6-12.

Niwa, K., Kobiyama, A., and Sakamoto, T. (2010). Interspecific hybridization in the haploid blade-forming marine crop *Porphyra* (Bangiales, Rhodophyta): occurrence of allodiploidy in surviving F_1 gametophytic blades. *J. Phycol.* 46: 693-702..

Niwa, K., and Sakamoto, T. (2010). Allopolyploidy in natural and cultivated populations of *Porphyra* (Bangiales, Rhodophyta). *J. Phycol.* in press.

Ohme, M., Kunifuji, Y., and Miura, A. (1986). Cross experiments of the color mutants in *Porphyra yezoensis* Ueda. *Jpn. J. Phycol.* 34: 101–106.

Ohme, M., and Miura, A. (1988). Tetrad analysis in conchospore germlings of *Porphyra yezoensis* (Rhodophyta, Bangiales). *Plant Sci.* 57: 135-140.

Park, E.-J., Fukuda, S., Endo, H., Kitade, Y., and Saga, N. (2007). Genetic polymorphism within *Porphyra yezoensis* (Bangiales, Rhodophyta) and related species from Japan and Korea detected by cleaved amplified polymorphic sequence analysis. *Eur. J. Phycol.* 42: 29-40.

Ryuto, H., Abe, T., Fukunishi, N., Kase, M., and Yano, Y. (2006). Heavy-ion beam irradiation system for biological samples in RIKEN. *J. Biomed. Nanotech.* 2: 88-93.

Shikazono, N., Suzuki, C., Kitamura, S., Watanabe, H., Tano, S., Tanaka, A. (2005). Analysis of mutations induced by carbon ions in *Arabidopsis thaliana. J. Exp. Bot.* 56: 587-596.

Sun, J., Liu, T., Guo, B., Jin, D., Weng, M., Feng, Y., Xu, P., Duan, D., Wang, B. (2006). Development of SSR primers from EST sequences and their application in germplasm identification of *Porphyra* lines (Rhodophyta). *Eur. J. Phycol.* 41: 329-336.

Suzuki, M., Watanabe, M., Kanai, T., Kase, Y., Yatagai, F., Kato, T., and Matsubara, S. (1996). LET dependence of cell death, mutation induction and chromatin damage in human cells irradiated with accelerated carbon ions. *Adv. Space Res.* 18: 127-136.

van der Meer J. P. (1990). Genetics. In K. Cole, and R. Sheath (Eds), *Biology of the red algae* (edition, pp. 103-122). Cambridge, UK: Cambridge University Press.

Waaland, J. R., Stiller, J. W., and Cheney, D. P. (2004). Macroalgal candidates for genomics. *J. Phycol.* 40: 26-33.

Wang, S. J., Zheng, Y. Z., Ma, L. B., Xu, P., and Zhu, J. Y. (2000). Gamma-rays induction of mutation in conchocelis of *Porphyra yezoensis*. *Chi. J. Oceanol. Limnol.* 18: 47-53.

Wang, J., Xu, P., Zhu, J., Zhou, W., Xu, J., Fei, X., and Zhang, X. (2008). The characterization of color mutations in Bangiaceae (Bangiales, Rhodophyta). *J. Appl. Phycol.* 20: 499–504.

Yan, X. H., and Aruga, Y. (1997). Induction of pigmentation mutants by treatment of monospore germlings with NNG in *Porphyra yezoensis* Ueda (Bangiales, Rhodophyta). *Algae* 12: 39-52.

Yan, X. H., and Aruga, Y. (2000). Genetic analysis of artificial pigmentation mutants in *Porphyra yezoensis* Ueda (Bangiales, Rhodophyta). *Phycol. Res.* 48: 177-187.

Yan, X. H., Fujita, Y., and Aruga, Y. (2000). Induction and characterization of pigmentation mutants in *Porphyra yezoensis* (Bangiales, Rhodophyta). *J. Appl. Phycol.* 12: 69-81.

Yan, X. H., Li, L., and Aruga, Y. (2005a). Genetic analysis of the position of meiosis in *Porphyra haitanensis* Chang et Zheng (Bangiales, Rhodophyta). *J. Appl. Phycol.* 17: 467-473.

Yan, X. H., Liang, Z. Q., Song, W. L., Huang, J., Ma, P., and Aruga, Y. (2005b). Induction and isolation of artificial pigmentation mutants in *Porphyra haitanensis* Chang et Zheng (Bangiales, Rhodophyta). *J. Fish. Chi.* 29: 166-172 (in Chinese with English abstract).

In: Porphyra yezoensis
Editor: Koji Mikami

ISBN 978-1-61122-304-0

Chapter 8

PROTOPLAST ISOLATION FROM *PORPHYRA* AND THE CELL WALL-DEGRADING ENZYMES

Toshiyoshi Araki* and Yoshiaki Umemoto†
Graduate School of Bioresources, Mie University, 1577 Kurimamachiya,
Tsu, Mie 514-8507, Japan

ABSTRACT

Porphyra (laver) is important edible seaweed cultivated widely in Japan, Korea, and China. The laver is the nutritious seaweed, which included functionality materials such as taurine, lutein, carotenoid, eicosapentaenoic acid, and unique polysaccharides (Yoshizawa et al.; 1995, Tao et al., 2008) as well as various vitamins and minerals. The laver is also expected as protein source because it includes about 40% protein per dried material—the same as the soy bean, which is called King of the plant proteins.

Since the British botanist Kathleen Mary Drew-Baker discovered in 1949, that *Conchocelis* was the diploid stage of *Porphyra* (Drew 1949), full-scale culture of laver began in Japan. After that, selective breeding of the laver was performed, and *Porphyra yezoensis* is cultivated widely now. However, in late years, marine environment turns worse with landfill or the shore protection works of near shore waters and gives laver culture the big damage by outbreak such as discoloration and red rot disease. In such a background, breeding of the laver of high quality having superior growth and disease resistance is expected by using cell fusion and a gene manipulation (Craigle, 1991; Minocha, 2001; Punja, 2001; Chawla, 2002). Establishment of the isolation method of the protoplasts is important for breeding by these biotechnologies. In addition, the protoplasts that raise digestion by removing a cell wall attract attention as new feed material of the cultured fish (Kalla et al., 2008a and 2008b; Khan et al., 2008).

However, the cell wall of the laver consists of β-1,4-mannan, β-1,3-xylan, and porphyran, so that protoplasts are not isolated from it with commercial cellulase and

* Correspondence should be addressed to: Tel: +81-59-231-9561; Fax: +81-59-231-9540, E-mail: araki@bio.mie-u.ac.jp. Laboratory for the Utilization of Aquatic Bioresources, Department of Life Science Graduate School of Bioresources, Mie University, 1577 Kurimamachiya, Tsu, Mie 514-8507, Japan.

† Laboratory for the Utilization of Aquatic Bioresources, Department of Life Science, Graduate School of Bioresources, Mie University, 1577 Kurimamachiya, Tsu, Mie 514-8507, Japan.

pectinase. Therefore, we tried to search bacteria that produce these cell wall-degrading enzymes such as β-1,4-mannanase, β-1,3-xylanase, and agarase capable of degrading porphyran, and characterized the properties of them. Subsequently, we examined a condition to isolate a large number of protoplasts from the laver with these enzymes. Finally, we observed a state of the cell wall reproduction of *Bangia* protoplasts by using carbohydrate binding module.

Keywords: *Porphyra*, protoplast, cell wall, polysaccharides, β-1,4-mannanase, β-1,3-xylanase, agarase, porphyran, carbohydrate-binding module, GFP, *Bangia atropurpurea*

ABBREVIATIONS

CBM, carbohydrate-binding module;
GFP, green fluorescent protein;
CFW, Calcofluor White;
PCR, polymerase chain reaction

1. ENZYMES CAPABLE OF DEGRADING THE CELL WALL OF *PORPHYRA*

The cell wall of the land plant consists of cellulose and hemicelluloses such as pectin and β-1,4-xylan mainly, whereas that of *Porphyra* consists of β-1,4-mannan and β-1,3-xylan as skeletal polysaccharides, and unique mucilage polysaccharide, porphyran (Frei and Preston, 1964; Rees, 1961). In this chapter, we describe screening of the cell wall-degrading bacteria from marine environment and characterization and properties of these enzymes.

1-1. β-1,4-Mannanase from *Vibrio* sp.MA-138

The polysaccharide, β-mannan is an important structural component of plant cell wall as well as cellulose and β-1,4-xylan. It is composed of β-1,4-linked D-mannose units. The polysaccharide is contained in some seaweeds such as Chlorophycea (Love and Percival, 1964) and Rhodophycea (Jones, 1950), in addition in terrestrial plant seeds such as coffee bean (Wolfrom et al., 1961) and ivory nut (Aspinall et al., 1953). It is supposed that the residues of these marine and terrestrial plants are decomposed by β-1,4-mannanase-producing microbes in nature. β-Mannanase (1,4-β-D-mannan mannano-hydrolase, EC 3.2.1.78), which catalyzes the random hydrolysis of β-1,4-mannosidic linkages in β-mannan, glucomannan, and galactomannan, is a useful tool for the structural analysis. There have been several papers on the purification and characterization of β-mannanases from bacteria (Duffaud et al., 1997; Akino et al., 1988; Emi et al., 1972; Oda et al., 1993; Sunna et al., 2001; Talbot and Sygusch, 1990; Yamaura et al., 1990; Tamaru et al., 1995), fungi (Takahashi et al., 1984), and higher plants (Villarroya and Petek, 1976). We also searched for β-mannanase-producing bacteria from marine environment (Araki and Kitamikado, 1978).

1-1-1. Screening and Identification of β-1,4-Mannanase-Producing Bacteria

β-Mannan was prepared from a green alga, *Codium fragile* (Love and Percival, 1964). The medium used for screening of bacteria is peptone agar medium (0.5% Polypepton, 0.1% yeast extract, 3% NaCl, 0.05% $MgSO_4$, 0.2% K_2HPO_4, and 0.04% KH_2PO_4, 1.5% agar, pH7.6) containing 1% β-mannan. Seventy-six samples were collected from the coastal sea areas of Mie, Wakayama, Osaka, and Kanagawa prefecture in Japan in 1990. Those included 21 from the bottom mud of coastal sea, 21 from seaweed, two from marine plankton, and 32 from intestinal contents of marine fish. A small quantity of each sample was added into 5 ml of sterile 3% NaCl solution. A loop of platinum from sample suspension was smeared on the peptone agar plate containing 1% β-mannan powder. Each plate was incubated for several days at 25°C. Colonies forming clear zoon around were isolated as mannanase-producing bacteria. The isolates cultured on slant media containing 0.5% konjac powder were stored in refrigerator and transferred monthly on homologous to maintain variability, or remained viable in storage at -80C for at least one year in the peptone medium after addition of glycerol to a final concentration of 30%.

As the result of screening, the four isolates capable of forming clear zone around the colonies on the plates containing β-mannan were obtained as β-mannanase-producing bacteria; three from seaweeds collected at coastal sea in Mie and one from bottom mud collected at coastal sea in Osaka. β-Mannanase activities were compared among these four isolates.

Cells taken from the fresh slant culture of each isolate were transferred to 20 ml lot of sterile peptone medium containing 0.5% β-mannan powder in a 100-ml cotton plugged flask. The flasks were incubated with shaking at 25°C for two days. Each culture was centrifuged at 9,000*g* for 20 min, and the supernatant was used as the crude enzyme solution. For measuring of β-mannanase activity, the reaction mixture consisted of 0.1 ml of enzyme solution, 0.4 ml of 2-(*N*-morpholino)ethanesulfonic acid (MES) buffer, pH 7.5, and 0.5 ml of 1% β-mannan suspension. After, these reaction mixtures were incubated at 37°C for 10 min and reducing sugar produced was determined by the Somogyi-Nelson method (Somogyi, 1952) and expressed as D-mannose. One unit of the β-mannanase activity was defined as the amount of enzyme producing the reducing sugar corresponding to 1.0 μmol of mannose from the substrate per min under the above conditions.

The bacterium having the highest β-mannanase activity among them was strain no. MA-138. Identification on genera of the isolate was made according to the diagram of Cowan and Steel, and the criteria described in Bergey's Manual of Systematic Bacteriology,Vol.1 (Baumann et al., 1984). The strain no. MA-138 belonged to *Vibrio*. The strain's characterization was as follows: Gram-negative, rod shaped, motility, polar flagella, non-sporangium, positive on catalase and oxidase tests, production of acid and no gas from glucose, fermentation on O-F test, and sensitivity on Vibriostatic reagent.

1-1-2. Purification and Characterization of β-1,4-Mannanase from Vibrio *sp.MA-138*

Cells of *Vibrio* sp.MA-138 were grown at 25°C for 24 h with shaking in 20 ml of a peptone medium containing 0.5% konjac powder as an inducer in a 100-ml Erlenmeyer flask and then were transferred to 1,000 ml of the same medium in a 3,000-ml Erlenmeyer flask. After two days incubation at 25°C, the culture was centrifuged at 9,000*g* at 4°C for 20 min. The supernatant was collected and used in the following steps. The active fraction from the culture supernatant at 75% saturation with solid ammonium sulfate was collected by centrifugation (9,000*g*, 30 min, 4°C), dissolved in a small volume of 50 mM Tris-HCl buffer, pH 7.3, and dialyzed against the same buffer. The dialyzed enzyme solution was purified by several chromatographic procedures including gel filtration, adsorption, and ion exchange chromatographies. One major active fraction and three minor active fractions were fractionated by the final Mono Q ion exchange chromatography. The major active fraction was purified to homogeneity on the basis of native-polyacrylamide gel electrophoresis (Native-PAGE). This purified enzyme was identified as a glycoprotein by periodic acid-Schiff (PAS) staining, and a monomeric protein with a molecular mass of 49 kDa by sodium dodecyl sulfate (SDS)-PAGE. The purified enzyme exhibited pI of 3.9, and maximal activity at pH 6.5 and 40°C, and the first 20 N-terminal amino acid sequence of the enzyme showed high homology with the N-terminal region of β-mannanase from *Streptomyces lividans* 66. The enzyme hydrolyzed at random the internal β-1,4-mannosidic linkages in β-mannan, glucomannan, and galactomannan to give various sizes of oligosaccharides. Hydrolysis products of the enzyme against β-mannan were mainly mannobiose (M2) and mannotriose (M3), in addition to a small amount of mannose (M1), mannotetraose (M4), -pentaose (M5), -hexaose (M6). It cleaved M4 to produce M3, M2 and M1. M5 was hydrolyzed in the same way to form either M3 and M2 or M4 and M1. The main products from M6 were M3 rather than M2 and M4, but the enzyme did not hydrolyze M2, M3 and *p*-nitrophenyl-β-D-mannoside. The major extracellular mannanase from *Vibrio* sp.MA-138 seems to hydrolyze preferentially the β-1,4-mannosidic linkages situated at the third position and followed by the fourth and second positions from the non-reducing end (Tamaru et al., 1995).

1-2. β-1,3-Xylanase from *Vibrio* sp.XY-214

β-1,3-Xylan is a polysaccharide constructed by polymerization of D-xylose units. Based on the structure, xylan can be divided into β-1,4-xylan and β-1,3-xylan. β-1,4-Xylan is found mainly in the hemicelluloses of the cell walls of land plants (Bastawde, 1992; Scheller and Ulvskov, 2010), while β-1,3-xylan constitutes the cell wall polysaccharides of seaweeds like *Caulerpa* and *Porphyra* (Iriki et al., 1960). Due to their wide industrial applications, β-1,4-xylan-degrading enzymes have been extensively studied (Hayashi et al., 1997; Wong et al., 1998; Beg et al., 2001). In contrast, β-1,3-xylan-degrading enzymes are not well studied. There have been few reports on endo-1,3-β-xylanases (EC 3.2.1.32), only three microorganisms, *Aspergillus terreus* A-07 (Chen et al., 1986), *Vibrio* sp.AX-4 (Aoki et al., 1988), and *Pseudomonas* sp.PT-5 (Yamaura et al., 1990). We also searched β-1,3-xylanase-producing bacteria from sea environment and isolated a bacterium strain no. XY-214, which has high β-1,3-xylanase productivity, from sea mud collected at Ise Bay in 1990 (Araki et al., 1999). The organism was found to belong to the genus *Vibrio*. The culture of the isolate was grown for two days at 25°C in 100-ml Erlenmeyer flasks containing 20 ml of the peptone medium, which consisted of (per liter) 5 g of Polypepton, 1 g of yeast extract, 30 g of NaCl, 5 g of $MgSO_4 \cdot 7H_2O$, 2 g of K_2HPO_4, and 0.4 g of KH_2PO_4 (pH 7.5), and 0.3% β-1,3-xylan as inducer. The seed culture was transferred to a five-liter fermenter containing three liters of the same medium, and then it was incubated for 48 h at 25°C with starring (200 rpm) and aeration (20 to 40 liters/min). After centrifugation (9,000*g* at 4°C for 20 min), the culture supernatant was treated with solid ammonium sulfate to a final concentration of 80% saturation. The mixture was stirred at 4°C and centrifuged at 9,000*g* for 30 min. The precipitate collected by centrifugation was dissolved in a small volume of 50 mM MES-NaOH buffer (pH 7.5) and dialyzed against the same buffer.

β-1,3-Xylan was prepared from a green seaweed; *Caurelpa racemosa* var. *laete-virens* and hydrolysis products (xylobiose to xylopentaose) of β-1,3-xylan with the β-1,3-xylanase were fractionated for preparing xylooligosaccharides by the method of the published papers (Iriki et al., 1960; Aoki et al., 1988). To measure enzyme activity, the enzyme solution (0.1 ml) was added to a mixture of 0.4 ml of 200 mM Mes-NaOH buffer (pH 7.5) and 0.5 ml of 1% β-1,3-xylan suspension. After incubating at 37°C for 10 min, the reducing sugar generated was measured by the Somogyi-Nelson method (Somogyi, 1952) and expressed as xylose. One unit of β-1,3-xylanase activity was defined as the amount of enzyme producing 1.0 μmol of xylose from the substrate per min under these conditions.

β-1,3-Xylanase from *Vibrio* sp.XY-214 was purified by several column chromatographies such as Q Sepharose FF, Ether Toyopearl 650S and Superdex 200. The final enzyme preparation was ascertained to be a single protein band by SDS-PAGE. After the electrophoresis was completed, another gel was placed on a 1.5% agarose gel sheet containing 0.1% glycol β-1,3-xylan and incubated for one hour at 37°C. The agarose gel sheet was stained with 1% Congo red solution. The single band was found to have the activity by zymogram. It was purified 83-fold from the culture fluid, and the recovery was 24%. The molecular mass estimated was 52 kDa by SDS-PAGE. Isoelectric point of the enzyme was 3.6 on isoelectric focusing. The optimal pH of the enzyme was observed at approximately 7.0 and lay in neutral side compared with those (pH 4.0-6.0) of fungus (Chen et al., 1986). The enzyme was stable between pH 5.0 to 8.0 during incubation at 4°C for 20h. When enzyme reaction was incubated for 10 min at various temperatures, the maximal activity of the enzyme was observed at around 37°C. The enzyme activity was stable under 30°C. For the measurement of the effects of several inhibitors, the enzyme was incubated for 10 min at 37°C with the chemical compounds (each at a final concentration of 1 mM) before the enzyme assay. The enzyme activity was inhibited completely by Cu^{2+} and Hg^{2+}, and over 90% by Ag^{+}, Zn^{2+}, and Pb^{2+}. The enzyme was 70 to 15% inhibited by Fe^{3+}, Al^{3+}, Mn^{2+}, dithiothreitol, and p-chloromercuribenzoic acid and 10 to 30% activated by Mg^{2+} and Ca^{2+}. It was also completely inhibited by 1 mM N-bromosuccinimide solution, which is the modification reagent for tryptophan residue in protein. It can, therefore, be presumed that tryptophan has an important role at the active site of β-1,3-xylanase.

β-1,3-Xylanase active band separated by SDS-PAGE was transferred onto a Clear Blot Membrane-p AE-6660 (Atto). The N-terminal amino acid sequence of the enzyme was analyzed by automated gas phase Edman degradation using a protein sequencer (model 476A; Applied Biosystems). The first 16 N-terminal amino acid sequence of the enzyme was analyzed. The first 13 N-terminal amino acid sequence of the *Vibrio* sp.XY-214 xylanase had 76.9 and 38.5% similarity with the same region of the enzymes of *Alcaligenes* sp.XY-234 (Araki et al., 1998a) and *Pseudomonas* sp.PT-5 (Yamaura et al., 1990), respectively.

The enzyme hydrolyzed β-1,3-xylan to produce mainly xylotriose and xylobiose, followed by a small amount of xylotetraose, xylose, xylopentaose, and xylohexaose. The enzyme had a weak action on xylotriose and produced a trace of xylose and xylobiose. The hydrolysis products of the enzyme from xylotetraose were xylotriose, xylobiose, and xylose, and those from xylopentaose were xylotetraose, xylotriose, xylobiose, and xylose. The enzyme did not act on xylobiose, p-nitrophenyl-β-D-xyloside (Sigma Co), β-1,4-xylan, curdlan, which is a polymer of the repeating units of β-1,3-linked D-glucose, and carboxymethylcellulose.

1-3. Agarases from *Vibrio* sp.PO-303

Porphyran is a unique sulfate polysaccharide, which is contained in the cell walls of Bangiacea like *Porphyra* and *Bangia*. It consists of three-linked β-**D**-galactose units, alternating with four-linked 3,6-anhydro-α-**L**-galactose units and also containing substituent groups such as sulfate. Agar is an abundant biopolymer found in agarophyte red algae as a component of cell walls. It is a complex molecule consisting of three-linked β-**D**-galactose units, alternating with four-linked 3,6-anhydro-α-**L**-galactose units (Rees, 1961; Percival and McDowell, 1967). Therefore, some agarases hydrolyze porphyran as well as agar. There have been reports on agarases from several bacteria and sea animals, including *Cytophaga* (Meulen and Harder, 1975), *Pseudomonas* (Morrice et al., 1983a, 1983b; Belas, 1989), *Streptomyces* (Bibb et al., 1987), *Vibrio* (Aoki et al., 1990a, 1990b; Sugano et al., 1993a, 1993b), *Alterococus agarolyticus* (Shieh et al., 1998), *Pseudoalteromonas* (Vera et al., 1998; Schroeder et al., 2003)、*Paenibacillus* (Hosoda et al., 2003) , and *Microbulbifer* (Ohta et al., 2004), *Zobellia* (Jam et al, 2005) as microbial sources, and *Littorina mandshurica* (Usov and Miroshnikova, 1975) as sea animal source.

Searched bacteria that produced agarase capable of cleaving porphyran from sea environment and isolated an excellent agarase-producing bacterium, strain no. PO-303 from sea mud collected from the coastal sea areas in Japan in 1991. The organism was found to belong to *Vibrio*. Porphyran was prepared from *Porphyra yezoensis* according to the method of Su and Hassid (Su and Hassid, 1962).

The organism was aerobically cultured in a five-liter, stirred tank fermenter containing three liters of the peptone medium supplemented 0.3% agar for at 25°C for 36h. The peptone medium for culturing was composed of 5g of peptone, 1g of yeast extract, 30*g* of NaCl, 5g of $MgSO_4 \cdot 7H_2O$, 2g of K_2HPO_4, and 0.4g of KH_2PO_4 per liter of distilled water (pH7.5). The culture was centrifuged at 9,000*g* at 4°C for 20 min, and the culture supernatant was treated with solid ammonium sulfate to a final concentration of 80% saturation. The mixture was stirred overnight at 4°C and centrifuged at 9,000*g* for 30 min. The precipitate was collected by centrifugation and dialyzing against 50 mM MES-NaOH buffer (pH 7.5). Enzyme assay was carried out as follows: a 100μl of enzyme sample was added into 900μl of 0.1M MES buffer (pH 7.5) containing 0.5% agar and then incubated at 38°C for 10 min. The reducing sugar content generated was determined by the Somogyi-Nelson method (Somogyi, 1952), with D-galactose as the standard. One unit of activity was defined as the amount of enzyme that produces 1μmol of D-galactose equivalents per min.

The dialyzed crude enzyme solution was fractionated by chromatographies using columns such as DEAE Toyopearl 650M, Gigapite, Ether-Toyopearl 650S, and Mono Q. By the purification procedure described above, three kinds of agarases (agarase-a, -b, -c) were purified, and each enzyme was characterized. The agarase-a, -b, and -c were purified 11.7, 44.2, and 32.3 folds with the recovery of 0.93, 2.09, and 3.75%, respectively. By SDS-PAGE, each final enzyme preparation gave a single-protein band. The molecular masses of the purified agarase-a, -b, and -c were estimated to be 87.5, 115, and 57.0 kDa, respectively, by SDS-PAGE, and isoelectric points of them were 6.6, 3.4, and 8.4, respectively. The pH optima for agarase-a and -c were 7.0 to 7.5, and that for agarase-b was 6.5. Agarase-a, -b, and -c were stable in pH range of 5.0 to 8.0, 4.0 to 10, and 5.0 to 9.0, respectively, at 4°C for 24h. Agarase-a, -b, and -c were most active at 45, 55, and 38°C, respectively. Relative activities of agarase-a, -b, and -c at 60°C were 25, 80, and 3%, respectively. Each enzyme retained full activity at temperature up to 40°C, but showed little activity above 60°C. Hg^{2+} completely inhibited these three enzymes. Ag^{+}, Cu^{2+}, and Zn^{2+} strongly inhibited agarase-a and -c more than agarase-b. Fe^{3+}, Al^{3+}, and EDTA completely or strongly inhibited only agarase-c, but gave no effect on agarase-a and -b.

The first 16 amino acids from the N-termini of agarase-a and the first 20 amino acids of agarase-b and -c were sequenced. In case of having been compared with amino acid sequences of the reported agarases, the sequence of agarase-b showed significant homology (75% identity in the first 20 amino acids) with the N-terminal region of agarase from *Vibrio* sp.JT0107. Agarase-a hydrolyzed agarose, agar, and porphyran to produce mainly neoagaro - tetraose and -hexaose, although it showed a weak action against porphyran. Most of the hydrolysis products of agarase-b from these polysaccharides were neoagarobiose, but a small amount of neoagarotetraose was also detected, especially from agarose. The hydrolysis products of agarase-c was mainly neoagaro-octaose and -decaose in addition to small quantity of neoagaro-tetraose and -hexaose from agarose, and predominantly neoagarooctaose in addition to neoagaro-tetraose, -hexaose, and -decaose from agar, but there was a little action on porphyran. Agarase-a hydrolyzed neoagarohexaose to produce neoagaro-biose and - tetraose, but did not neoagaro-biose and -tetraose. The hydrolysis products of agarase-b from neoagaro-tetraose and -hexaose were predominantly neoagarobiose. Agarase-c could not cleave neoagaro-biose, -tetraose, –hexaose, and -octaose (Table 1). Agarase-a, -b, and -c did not act on κ-, λ-, and ι-carrageenan, and *p*-nitrophenyl α- and β-**D**-galactosides. All these enzymes were endo-type agarases, and agarase-c was a new enzyme (Araki et al., 1998b; Dong et al., 2006a).

2. Isolation of Protoplasts from *Porphyra*

Protoplast is the living material including the protoplasm and plasma membrane after the cell wall has been removed from plants, bacteria, and fungi, etc., and useful tool for breeding by cell fusion and gene manipulation. Protoplasts were isolated for the first time by microsurgery on plasmolyzed cells by Klercker in 1892, using mechanical method (Sateesh, 2003), though the yields were extremely low and this method was not useful. In 1960, Cocking used cellulase from a fungus *Myrothecium verrucaria* for protoplast isolation from tomato root tips (Cocking, 1966). However, there was little work for the next ten years until the cell wall-degradation enzymes were sold commercially. Takebe et al. established the useful method of protoplast isolation by treating plant tissues with commercial enzymes like cellulase and pectinase et al (Takebe et al., 1968).

Table 1. Characterization of several agarases from bacteria

		Optimal pH	Optimal Temp. (°C)	Molecular Mass (kDa)	Agar	NA4	NA6	NA8
Vibrio sp. PO-303	a	7.0-7.5	45	87.5	6, 4	-	4, 2	4, 2
	b	6.5	55	115	2, 4	2	2	2
	c	7.0-7.5	38	57	8,10 > 6,4	-	-	-
P. atlantica	I	7.0	ND	32	4, 6, 2	-	4, 2	ND
	II	7.0	ND	ND	2, 4	2	2	ND
Vibrio sp. AP-2	a	6.5	45	34	4, 6, 8	-	2, 4	2, 4
	b	5.5	45	20	2 ,4	2	2	2
	c	7.0	45	18	4, 6, 8	-	2, 4	2, 4
Pseudonomas PT-5		8.5	ND	31	ND	ND	ND	ND
Alteromonas C-1		6.5	30	52	ND	ND	ND	ND
Vibrio sp. JT0107	A	8.0	30	107	ND	2	ND	ND
Pseudomonas like	I	6.7	38	127+127	ND	-	4, 2	2, 4, 6
	IIb	6.7	43	64	ND	-	4, 2	4
Cytophaga fluvensis		6.3	35	26.5	ND	ND	4, 2	ND

-: negligible activity. ND: not determined.
NA2, NA4, NA6, and NA8: Neoagaro-biose, -tetraose, -hexaose, and -octaose.
Numbers 2, 4, and 6 in columns 5, 6, 7, and 8 show neoagaro-biose, -tetraose, and -hexaose.

Protoplast isolation from seaweeds was published in 1984, by two groups. Saga et al. succeeded to isolate protoplasts from a brown alga *Laminaria* and a red alga *Porphyra* by using the crude enzyme extracted from sea urchin (Saga, 1984; Saga and Sakai, 1984). Polne-Fuller et al. isolated protoplasts from *Porphyra perforata* by the crude enzyme from abalone (Polne-Fuller and Gibor, 1984). At first, protoplast isolation from seaweed has been conducted by using the digestive enzymes from marine herbivores such as abalone, turbo cornat, and sea urchin, etc., which are animals that are adapted to eat seaweed. These methods had the fault that unevenness was big among the enzyme lots. Afterwards, the method was improved by using the enzyme mixture prepared by addition of commercial enzymes like cellulase and pectinase into the digestive enzymes of marine herbivores.

However, the cell walls of seaweed are composed of several kinds of polysaccharides. The cell wall compositions of brown algae are cellulose, alginate, and fucoidan. Those of Florideophyceae are cellulose and agar or cellulose and carageenan. In Chlorophyceae, the cell walls of Ulvales, Codiales, and Caulerpales consisted of cellulose and sulfated glucuronoxylorhamnan, β-mannan and sulfated xyloarabinogalactan, and β-1,3-xylan and sulfated xyloarabinogalactan, respectively. The cell walls of Bangiophyceae that *Bangia* and *Porphyra* belong are composed of β-mannan, β-1,3-xylan, and porphyran. Therefore, we cannot isolate protoplasts with commercial cellulase and pectinase. The reason why protoplasts were able to be isolated from *Porphyra* with the digestive enzymes of marine harbivers is that β-mannanase, β-1,3-xylanase, and agarase were contained in them. However, the cell wall-degrading enzyme activities are not so strong, and the content ratio of each enzyme is different among the sources so that the instability of the enzyme activities disturbs the routine of the protoplast isolation. There are some papers on the synergistic effect for protoplast isolation by blending the digestive enzymes of marine herbivores with commercial cellulase and pectinase, because some commercial enzymes contain β-mannanase activities, and slight β-1,3-xylanase and agarase activities. Fujita and Migita prepared the crude enzyme from the cultured supernatant of a marine bacterium *Pseudomonas* sp. strain P-1, which causes "Green spot rotting" on *Porphyra* thalli. The strain P-1 produced the cell wall-degrading enzymes in the culture fluid by being cultured in the liquid medium containing 1% *Porphyra* powder as an inducer. They have reported on protoplast isolation from *Porphyra* by using the bacterial crude enzyme solution (Fujita and Migita, 1985). However, the bacterial enzyme solution is mixture of the cell wall-degrading enzymes and the rate of three enzymes contained in the solution is not clear.

We developed a method to prepare a large quantity of the three kinds of enzymes from marine bacteria, which are required for protoplast isolation from *Porphyra*, as described in paragraph 1. We are able to arrange the mixture ratio of each enzyme freely by using these three kinds of enzymes so that will isolate protoplasts efficiently. In this paragraph, we describe the several conditions for protoplast isolation with these enzymes (Araki et al., 1987; Araki and Morishita, 1990).

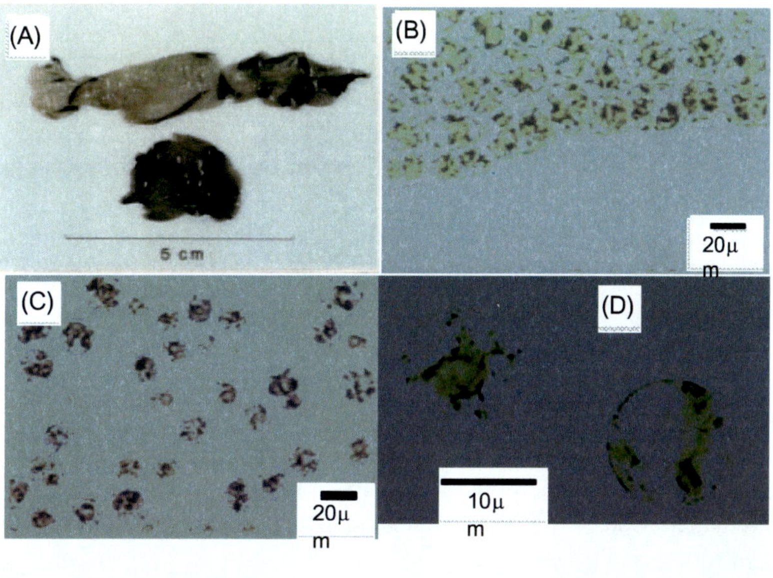

Figure 1. Isolation of protoplasts from *Porphyra* yezoensis by using bacterial enzymes. The thalli of *P. yezoensis* pretreated with 5% papain were incubated for 90 min at 22°C in the enzyme mixture containing one unit each of β-1,4-mannanase,β-1,3-xylanase and β-agarase, and 0.7 M mannitol in 20 mM MES buffer, pH 7.5. (A) Thalli of *P. yezoensis*, (B) The surface of the thallus, (C) and (D) Protoplasts.

2-1. Isolation of Protoplasts

For protoplast isolation, the thalli of *Porphyra yezoensis* were cultured in ASP_{12}(NTA) medium (Provasoli et al., 1957) at 17°C under a 9:15 h of light and dark cycle with agitation (30 strokes · min^{-1}) on a reciprocal shaker. A white fluorescent light was adjusted to 48 μ mol photons · m^{-2} · S^{-1} on the outer surface of the Petri dishes, which were 9-cm in diameter and contained 40 ml of the medium.

The thalli of *P. yezoensis* were immersed in 30 ml of papain solution (ASP_{12}(NTA) containing 5% papain and 0.7 M mannitol) in a 100-ml beaker, and incubated at 22°C for 30 min with agitation. The thalli were washed several times with ASP_{12}(NTA) medium containing 0.7 M mannitol and finely chopped using a razor. The fragments (approximately 200 mg fresh weight) were then placed into a 50-ml Erlenmeyer flask containing 8 ml of an enzyme mixture consisting of one unit each of β-mannanase, β-1,3-xylanase and β-agarase, and 0.7 M mannitol in ASP_{12}(NTA) medium. Following incubation with gentle agitation for 90 min at 22°C, the reaction mixture was filtered through a 40-μm nylon mesh. Protoplasts were collected by centrifugation at 1,000*g* for five min, and the resulting pellet was gently re-suspended in 6 ml of ASP_{12}(NTA) medium containing 0.7 M mannitol. As shown in Figure 1, a lot of protoplasts were isolated.

2-2. Effect of Pretreatment of Protease on Protoplast Isolation

The thalli of *P. yezoensis* were agitated in the papain solution containing different concentration of papain for 30 min at 22°C and washed several times with ASP_{12}(NTA) containing 0.7 M mannitol. After the thalli were cut with a razor the fragments (about 50 mg in wet weight) were dipped into the cell wall-digestive enzyme solution containing one unit each of β-mannanase, β-1,3-xylanase, and agarase and then incubated with agitation for 90 min at 22°C. The released protoplasts were counted so that 5% papain pretreatment was most effective. Protoplast yield obtained by treatment with 5% papain increased three times compared to that without papain treatment. The survival rate of those protoplasts was about 80% after one-day culture.

It has been reported that protease pretreatment made the yield of protoplasts increase by some groups (Mizukami et al., 1992; Fujita and Saito, 1990; Gall et al., 1993).

2-3. Effects of pH, Temperature, Osmoticum, Enzyme Concentration, and Reaction Time

We examined several conditions for protoplast isolation from the thalli of *P. yezoensis* by using bacterial enzymes. The tests were conducted by using the thalli after 5% papain pretreatment.

1) pH

The pH of reaction mixture for protoplast isolation is affected greatly by optimal pH of the enzymes. We prepared two kinds of the enzyme mixtures. One is the enzyme mixture-A

(MAX-A) that contains β-mannanase from *Aeromonas* sp.F-25, agarase from *Vibrio* sp.AP-2, and β-1,3-xylanase from *Vibrio* sp. AX-4. The other is the enzyme mixture-B (MAX-B) that contains β-mannanase from *Vibrio sp.*MA-138, agarase from *Vibrio* sp.PO-303, and β-1,3-xylanase from *Vibrio* sp.XY-214. When the thalli were treated with MAX-A, pH 6.0 was higher in the yield of protoplasts 1.7 times than pH 7.5. On the other hand, pH 7.5 exhibited high isolation quantity about two times that of pH 6.0 when treated with MAX-B. The reason why is that the optimal pH of each enzyme in MAX-A is around pH 6.0 and that of each enzyme in MAX-B is around pH 7.5. When protoplasts are isolated by using cellulase from *Trichoderma* and pectinase from *Rhizopus*, the enzyme reaction is generally conducted at pH 5.0 to 6.0, because the optimal pH of former is 4.0 to 5.0 against cellulose, and that of latter is 5.0 to 6.0 against pectin. However, the enzyme reaction is preferable at the pH condition of weak alkali, when protoplasts are especially used for regeneration, because sea environment where *Porphyra* has habitation is around pH 8.0.

2) Reaction temperature

Effect of different temperatures (17, 22, 27, 32, and 37°C) was examined on protoplast isolation from *P. yezoensis*. High yield of protoplasts was obtained at 22, 27, and 32°C. However, the yields at 17 and 37°C were 55.6 and 9.3% compared to the suitable temperatures, respectively. The reason why is that the activities of the enzymes are low at 17°C, while 37°C is good condition for their activities but are too high as habitable temperature of the released protoplasts. Therefore, we usually perform enzyme reaction at 22.

3) Comparison of diverse carbohydrates as osmoticum in the isolation of protoplasts

Isolation and regeneration of protoplasts require osmoticum in the culture liquid, because they are easily affected by the osmotic pressure of the biotope. We examined the effect of different osmoticums, and sorbitol was found to be the best as osmotic pressure agent for protoplast isolation compared to mannitol, glycerol, glucose, and sucrose. When the protoplast yield obtained by treatment with sorbitol was set as 100, those treated with mannitol, glycerol, glucose, and sucrose were 99, 77, 53, and 37, respectively. This result supports the validity that sorbitol and mannitol are used generally as osmoticum. These agents are commonly used at the density of 0.5 to 1.2M.

4) Enzyme concentration and reaction time

The thalli (about 200 mg in wet weight) pretreated with papain were dipped into the cell wall-digestive enzyme solution (8 ml) containing 0.5, 1.0, and 2.0 unit each of β-mannanase, β-1,3-xylanase, and agarase, and incubated under agitation for 60, 120, 180 min at 22°C. When the thalli were treated with two units each of β-mannanase, β-1,3-xylanase, and agarase, 2.6, 4.1, and 5.6 x 10^6 of protoplasts were released after 60 -, 120-, and 180-min incubations, respectively. The number of protoplasts released from the thalli increased in proportion to the concentration of the cell wall-degrading enzymes and incubation time. Therefore, we usually treat the thalli (200 mg in wet weight) with the enzyme mixture solution (8 ml) containing five units of β-mannanase, two units of β-1,3-xylanase, and two units of agarase, and 0.7 M mannitol, at 22°C for 90 min and obtain 10^7 order of protoplasts from 1g of the thalli.

Although the digestive enzymes of marine herbivores such as a sea urchin, an abalone, and a turban shell are often used for a protoplast isolation from *Porphyra*, there have been few reports on hydrolysis activities toward β-mannan, β-1,3-xylan, and porphyran of their enzymes.

Table 2. Activities of polysacharide-degrading enzymes and protease in viscera of marineanimals (units/wet weight g)

Origins Substrates	Fish *Siganus fuscescens*	*Navodon modestus*	*Halichoeres poecilopterus*	*Chrysophrys major*	Shells *Anadara broughtonii*	*Solecurtus wilsoni*	*Ostrea gigas*
Agar	0.039	0.016	0.086	0.007	0.004	0.104	0.008
β-1,4-Mannan	-	0.004	0.013	-	0.053	0.163	-
Porphyran	0.107	0.102	0.109	0.065	0.047	0.087	-
β -1,3-Xylan	-	-	0.007	-	-	-	-
Casein	0.309	0.280	0.124	0.573	0.206	0.289	0.218

Origins Substrates	Shells *Hariotis discus*	*Turbo cornutus*	*Hemifusus ternatanus*	*Trochus maculatus*	*Babyloma japonica*	Sea urchin *Anthocidariu crasisspina*	Crab *Eriocheir japonicus*
Agar	0.032	0.171	0.082	0.109	0.029	0.031	0.092
β -1,4-Mannan	0.204	0.529	-	0.277	0.070	0.004	0.484
Porphyran	0.110	0.079	0.101	0.041	0.136	0.006	0.023
β -1,3-Xylan	-	0.106	0.005	0.117	0.002	-	0.009
Casein	0.873	0.018	1.084	0.345	3.271	0.072	0.147

negligible activity.

Table 3. Activities of polysacharide-degrading enzymes and protease of the acetone powder of shell hepatopancreas and the microbe enzyme preparations (units/dry weight g)

Origins Substrate	Preparation Shells Topshell acetone powder	Enzymes obtained Abalone acetone powder	commercially Microbial enzymes Cellulase ONOZUKA-RS	 Macerozyme R-10	 Driselase
Agar	0.153	0.153	0.200	-	-
β-1,4-Mannan	1.201	0.839	8.294	2.523	6.339
Porphyran	0.269	0.087	0.056	-	-
β-1,3-Xylan	0.071	-	0.174	-	0.349
Casein	0.078	-	38.88	-	9.474

-: negligible activity.

Yamaguchi et al. measured the polysaccharides-hydrolyzing activities and protease activity of the digestive enzymes from four kinds of seaweed-eating fish, eight kinds of shellfish, a sea urchin, and a crustacean (Yamaguchi et al., 1989). As shown in Table 2, their enzyme activities toward the *Porphyra* polysaccharides were very low, because the activities of β-mannanase, β-1,3-xylanase, agarase, and porphyranase per 1 g of digestive organs of each specimen were 0-0.529, 0-0.117, 0.004-0.171, and 0-0.136 units, respectively. Table 3 shows the enzyme activities of the acetone powder of the midgut glant extract of turban shell and of commercial abalone acetone powder, Cellulase ONOZUKA-RS, Macerozyme R-10, and Driselase. The β-mannanase activities of Cellulase ONOZUKA-RS and Driselase were comparatively high and showed to be 8.3 and 6.3 units per 1g of the dry powder, respectively, but their agarase, β-1,3-xylanase, and porphyranase activities were not so strong. Saga succeeded to isolate protoplasts from a green alga *Enteromorpha* by using Cellulase ONOZUKA-RS and Macerozyme R-10, but was not able to isolate them from *Porphyra* (Saga 1984). The reason why is thought to be that these enzymes having strong cellulase activity affected to remove the cell wall of *Enteromorpha,* which consists of cellulose, but were not able to isolate protoplasts from *Porphyra,* because their β-1,3-xylanase and porphyranase activities were very weak. Gall reported to be able to isolate protoplasts by using the enzyme mixture of Cellulase ONOZUKA-RS and agarase (two units) (Gall et al., 1993). As Fujita et al. have isolated effectively protoplasts from *P. yezoensis* with the culture supernatant of *Pseudomonas* sp.P-1, it is supposed to contain the three kinds of the cell wall-degrading enzymes (Fujita and Migita, 1985). However, they do not publish on the enzyme activities of the bacterium culture supernatant.

As described in a paragraph 1, we measured the enzyme activities of the culture supernatants of three kinds of bacteria isolated from sea environment. As a result, 591 units of β-mannanase, 93 units of β-1,3-xylanase, and 517 units of agarase were yielded from the culture supernatants (1 L) of *Vibrio* sp.MA-138, *Vibrio* sp.XY-214, and *Vibrio* sp.PO-303, respectively. In case of bacteria, the quantity of the enzyme production is in proportion to the volume of the culture liquid so that we can prepare a large quantity of enzyme by a large-scale culture. Each the cell wall-degrading enzyme is able to be prepared in a large quantity effectively by using three different bacteria strains. Therefore, microorganisms like bacteria have a big advantage as an enzyme source.

Recently, Kamei et al. cloned the β-mannanase-, β-1,3-xylanase-, and porphyranase-encoding genes from the genomic library of *Pseudomonas* sp.ND137 and expressed in *Escherichia* coli. They isolate 1.4 x 10^4 cells/mg wet weight of the thalli of *Porphyra yezoensis* by using these recombinant enzymes (Aoki and Kamei, 2006). We also cloned the genes of the cell wall-degrading enzymes from the genomic DNA of isolate; two β-mannanase genes (*manA* and *manC*) from the genomic DNA of *Vibrio* sp.MA-138 (Tamaru et al., 1997; Tanaka et al., 2009), one β-1,3-xylanase from *Vibrio* sp.XY-214 (Araki et al., 2000), and five agarase genes (*agaA, agaB, agaC, agaD,* and *AgaE*) from *Vibrio* sp.PO-303 (Dong J et al., 2006a, 2006b; Dong et al., 2007) and determined their sequences. We ascertained to be able to isolate a large number of protoplasts from the thalli of *Porphyra yezoensis* by using these recombinant enzymes expressed in *Escherichia coli.*

As mentioned above, the enzyme of the microbe origins such as bacteria will come to be used for protoplast isolation from *Porphyra* from now. In addition, it is expected to be constructed the production system of strong recombinant enzymes by a gene manipulation using the enzyme genes of not only microbes but also animals and plants. Now, β-mannanase, β-1,3-xylanase, and agarase of the bacteria origin are marketed by Yakult Yakuhin Kogyo Co., LTD in Japan.

3. Observation of the Cell Wall Regeneration in *Bangia Atropurpurea* Protoplasts by Using a Mannan-Specific Carbohydrate Binding Module

Bangia atropurpurea is a red alga that belongs in Bangiacea, Bangiales, Protoflorideophycidae as well as Porphyra. The cell wall of the alga consists of β-mannan, porphyran and β-1,3-xylan so that protoplasts are isolated from the fronds by using Porphyra cell wall-degrading enzymes. As the fronds of *B. atropurpurea* are hair-like form, protoplast isolation from the fronds is easily compared to thalli of *Porphyra* that are flat form. *B. atropurpurea* strain no. FA-90 cultured in our laboratory is sterile and has been maintained all year round by repetitive asexual reproduction, which cultures repeatedly monospores and fronds (Araki et al., 1994). Therefore, the *B. atropurpurea* has a very superior property as model alga on studying the physiology and biochemistry of seaweeds. As a part of fundamental research on seaweed, we tried to observe the cell wall regeneration of *Bangia* protoplasts by using the fusion fluorescent dye protein of GFP and CBM.

Generally, polysaccharide-degrading enzymes such as cellulases (β-1,4-glucanases) and β-1,4-xylanases possess a modular structure, consisting of a catalytic module and one or more non-catalytic CBMs (carbohydrate-binding modules) (Coutinho and Henrissat, 1999; Okazaki et al., 2002; Doi and Kosugi, 2004; Bayer et al., 2004; Doi, 2008). On the basis of the similarity of their amino acid sequences, GHs (glycoside hydrolases) and CBMs are now classified into 118 and 59 families, respectively (Current information on GHs and CBMs is available at the CAZy website: http://afmb.cnrs-mrs.fr/CAZy/) (Henrissat, 1991; Henrissat and Bairoch, 1996). This classification system divides β-mannanases into two families, GH5 and 26. Although β-mannanases have been prepared from bacteria, fungi, and plants, and many of the genes for the enzymes have been cloned and sequenced, so far there have been no reports on β-mannanases from *Vibrio* sp. We found that a marine bacterium *Vibrio* sp. strain MA-138 secretes multiple β-mannanases into the growth medium in the presence of an inducer such as β-mannan, konjac powder, or mannose (Tamaru et al., 1995). Two mannanase genes (manA and manC) were cloned, sequenced, and expressed as recombinant mannanases (ManA and ManC) in *Escherichia coli*. As a result of analysis, ManC was found to contain CBM in C-terminal addition to catalytic module.

In this paragraph, we describe “Cloning and sequence of CBM gene in ManC gene from genomic DNA of *Vibrio* sp.MA-138” and “Visualization of the regenerating cell walls by using a fused green fluorescent protein GFP-CBM.”

3-1. Cloning and Sequence of CBM Gene in ManC Gene from Genomic DNA of *Vibrio* sp.MA-138

Genomic DNA of *Vibrio* sp. strain MA-138 was partially digested with *Eco*RI and electrophoresed on a 0.7% agarose gel. Fragments in the range of 4 to 10 kbp were excised from the gel and purified with a Wizard gel and PCR clean-up system (Promega). The purified fragments were ligated into the dephosphorylated *Eco*RI site of λZAPII. The ligation mixture was packaged *in vitro* with GigapackIII Gold (Stratagene La Jolla, CA, USA). Bacteriophage λZAPII was mixed with *E. coli* XL1-Blue MRF′ and was incubated on NZY agar plates containing 1 mM isopropyl-β-D-thiogalactopyranoside (IPTG) to yield plaques. Plates were incubated overnight at 37°C to obtain visible plaques, and 1.0% agar solution containing 0.15% konjac glucomannan was overlaid to screen the plaques having mannanase activity. After overnight incubation, one plaque having mannanase activity was recognized by the formation of a clear halo on a red background after staining with 1% Congo red and de-staining with 1 M NaCl. The positive plaque isolated was subjected to *in vivo* excision to excise the pBluescript SK(−) phagemid containing the cloned insert from the λZAPII vector with ExAssist helper phage (Stratagene). The obtained plasmid was named pManC, and transfected into *E. coli* SOLR. Positive colonies were selected by activity staining with Congo red. The DNA insert in the pManC vector was sequenced, and the complete *man5C* gene consisted of 2,010 bp nucleotides encoding a protein of 669 amino acids with a molecular weight of 76,307 and was found to contain a signal peptide consisting of 26 amino acids from the first methionine (Shine and Dalgarno, 1975).

We constructed two derivatives such as pMan5C/CM (residues Leu23 to Ala348) that encodes the catalytic module belonging to the glycoside hydrolase family 5 and pMan5C/CBM (residues 493 to 669) that encodes the family 27 CBM. Each *E. coli* BL21 transformant carrying pMan5C/CM or pMan5C/CBM was cultivated at 37°C in 800 ml of LB medium in the presence of ampicillin (100 mg ml^{-1}). When the optical density at 600 nm of the culture reached about 0.5, the temperature was lowered to 15°C, and IPTG was added to the culture to give a final concentration of 1 mM for induction of gene expression. After an additional incubation of 20 h at 15°C, the cells harvested by centrifugation were suspended in a buffer and disrupted on ice by sonication. Each protein of rMan5C/CM or rMan5C/CBM contained in the cell lysate was collected by centrifugation and purified by a HiTrap chelating HP column (GE Healthcare, UK Ltd) precharged with Ni^{2+} and a Resource Q column (GE Healthcare) according to the manufacturer's instructions. The final preparations of purified rMan5C/CM and rMan5C/CBM were ascertained to be homogeneous by SDS-PAGE.

We analyzed β-mannan-binding abilities of rMan5C/CM and rMan5C/CBM. As a result, both proteins did not show any binding abilities to insoluble polysaccharides, such as codium mannan, Avicel, and β-1,3-xylan. The affinities of rMan5C/CM and Man5C/CBM for series of different soluble polysaccharides were also qualitatively evaluated by native affinity gel electrophoresis. In the case of rMan5C/CBM, the migration of proteins was significantly retarded by the inclusion of glycol β-1,4-mannan, konjac glucomannan, and guar gum galactomannan, but the retardation was not observed in case of rMan5C/CM. These results indicate that a putative CBM27 region of Man5C is a CBM, which has an ability to bind to soluble β-mannans.

3-2. Construction of the Fusion Protein of GFP and CBM27

The fusion protein of GFP and CBM27 (GFP-CBM27) was constructed as described below. The GFP gene was amplified from pRSET/EmGFP (Invitrogen Co., Carlsbad, California, U.S.A.) as a template by PCR. The gene encoding a region of CBM27, including the linker sequence (residues 466 to 669), was amplified from *man5C*/pCold vector that contained the β-mannanase gene (*man5C*) of *Vibrio* sp. strain MA-138 (Tanaka et al., 2009), as a template by PCR. The amplified GFP and CBM27 gene fragments that attached adaptor sequences were digested with *Eco*RI, and these two fragments were ligated to each other using DNA Ligation Kit Ver. 2.1 (Takara). To amplify the DNA fragment encoding the GFP-CBM27 fusion protein, PCR was performed using the reaction mixture as a template. The PCR product was digested with *Sac*I and *Xho*I, and ligated into a pCold TF DNA vector, linearized with the same enzymes to construct GFP/CBM27/pCold. Furthermore, GFP/pCold vector was also constructed for expression of normal GFP, as described below. After the amplification of the GFP gene, the resulting PCR product was digested with *Sac*I and *Eco*RI, and ligated into a pCold TF DNA vector, linearized using the same enzymes. The absence of undesired mutation in the amplified DNA fragments was verified by DNA sequencing. The plasmid GFP/CBM27/pCold vector, which provides GFP-CBM27 fused with a six-His tag and a trigger factor (TF) at the N-terminus, was transformed into *E. coli* BL21 (DE3)-competent cells and used for the production of the recombinant protein. TF is a type of chaperone in *E. coli*, and it functions as a tag for promoting the solubility of polypeptides.

The *E. coli* BL21 (DE3) transformant carrying the GFP/CBM27/pCold vector was cultivated at 37°C in 800 ml of LB medium containing ampicillin (50 $mg \cdot ml^{-1}$). When the optical density of the culture reached approximately 0.5 at 600 nm, the temperature was changed to 15°C, and isopropyl-β-D-galactopyranoside was added to the culture at a final concentration of 1 mM for the induction of gene expression. After an additional incubation for 20 h at 15°C, the cells harvested by centrifugation were suspended in 30 ml of 50 mM 2-morpholinoethanesulfonic acid (MES)-NaOH buffer (pH 7.0), and disrupted on ice by sonication. After centrifugation, the supernatant of the cell lysate was collected and used as the GFP-CBM27 fusion protein in all the experiments in this study (Figure 2). To serve as a control in the binding assay, GFP alone was also expressed from the *E. coli* BL21 (DE3) transformant harboring GFP/pCold vector by the same procedure.

3-3. Isolation and Regeneration of *Bangia Atropurpurea* Protoplasts

The fronds of a red alga, *B. atropurpurea* strain no. FA-90, were collected from the mouth of a river in Fukuoka Prefecture in Japan on January in 1990 (Araki et al., 1994). The fronds were cultured in ASP_{12}(NTA) medium of Provasoli at 17 °C, 9hL : 15hD (48 $\mu Em^{-2}s^{-1}$). The strain no. FA-90 is sterile and has been maintained all year round by repetitive asexual reproduction, which cultures repeatedly monospores and fronds. Unless otherwise noted, the fronds of 3 to 5 cm in length were used for the isolation of protoplasts.

The protoplasts were isolated from the fronds of *B. atropurpurea* by using *Porphyra* cell wall-degrading enzymes.

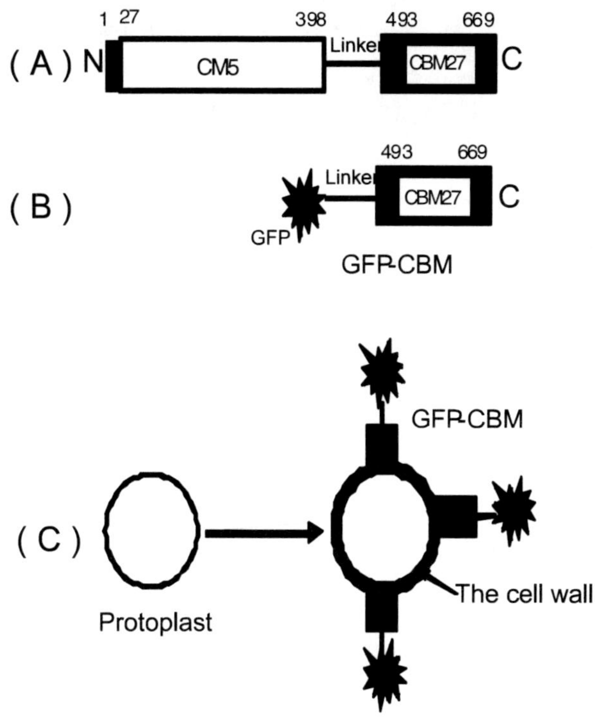

Figure 2. Schematic representation of the fusion protein of GFP-CBM27, which links to a regenerating cell wall of protoplast.(A) The modular architecture of Man5C from *Vibrio* sp.MA-138. Man5C contains a signal peptide, a glycoside hydrolase family-5 (GH5) catalytic module, linker region, and a CBM27, in this order from the N terminus. (B) The fusion protein GFP-CBM27 contains a GFP, a linker region, and a CBM27 derived from Man5C, in this order from N terminus. (C) linkage of GFP-CBM27 to a generating cell wall of protoplast.

The branches excised from the *B. atropurpurea* fronds were immersed in 30 ml of papain solution (20 mM MES buffer, pH 7.5 containing 5% papain and 0.5 M mannitol) in a 100-ml beaker, and incubated at 22°C for 30 min with agitation. The branches were washed several times with 20 mM MES buffer, pH 7.5 containing 0.5 M mannitol and finely chopped using a razor. The fragments (approximately 200 mg fresh weight) were then placed into a 50-ml Erlenmeyer flask containing 8 ml of an enzyme mixture consisting of five units of β-mannanase, two units of β-agarase, two units of β-1,3-xylanase, and 0.5 M mannitol in 20 mM MES buffer, pH 7.5. Following incubation with gentle agitation for 90 min at 22°C, the reaction mixture was filtered through a 40-μm nylon mesh. Protoplasts were collected by centrifugation at 1,000*g* for five min, and the resulting pellet was gently re-suspended in 6 ml of ASP_{12}(NTA) medium containing 0.5 M mannitol. This was centrifuged again under the same conditions, and the resultant pellet was gently re-suspended in 6 ml of ASP_{12}(NTA) medium containing 0.25 M mannitol. The presence of residual mannitol in the protoplast culture medium often causes undesired microbial growth. Therefore, by repeating this procedure of centrifugation and re-suspension, the concentration of mannitol was gradually reduced, and the protoplasts were finally re-suspended in 6 ml of ASP_{12}(NTA) medium without mannitol. After the complete removal of cell wall was confirmed by staining with 0.01% (w/v) CFW (Maeda and Ishida, 1967), some protoplasts were stored in sterilized test tubes for 12 h at 17°C in the dark. The culture was then incubated under a white fluorescent light of approximately 48 μmol photons · m^{-2} · s^{-1} in a 9L:15D cycle at 17°C. The other protoplasts were pipetted on to cover glasses (18 x 18 mm, Matsunami) in 9-cm Petri dishes containing 30 ml of ASP_{12}(NTA) medium and cultured under the same conditions as described above.

3-4. Visualization of β-1.4-Mannan in the Regenerating Cell Wall of *B. Atropurpurea* Protoplasts

After the isolation of protoplasts, we labeled them with the GFP-CBM27 fusion protein during their regeneration processes and analyzed the assembly and distribution of β-mannan, which is a major component of the cell wall polysaccharides in *B. atropurpurea*, by fluorescence microscopy. Immediately after protoplast isolation, only red autofluorescence of the photosynthetic pigments was observed, and no fluorescence was observed in GFP-CBM27-labeled or CFW-stained protoplasts. This was an indication of the complete removal of the cell wall. After culturing, the binding of GFP-CBM27 to the regenerating cell walls was observed but GFP used alone as the control did not bind to the cell walls at any time during the protoplast regeneration process.

After 12 h of culture, approximately 3% of the protoplasts began to form β-mannan at certain areas of their cell surface. As the protoplast culture progressed, the protoplasts revealed fluorescence on the entire surface. The percentages of cells emitting green fluorescence on their cell surfaces were 0, 3, 12, 17, 29, and 25% after 0, 12, 24, 36, 48, and 60 h of culture in test tube, respectively. Some of the protoplasts cultured on the cover glasses in the Petri dishes containing ASP_{12}(NTA) medium began to form rhizoids after three days of culture. During this culture period, the GFP-CBM27 fusion protein was able to bind the regenerated cell walls of almost all the cells, and a strong fluorescence was observed (Figure3). The first cell divisions were observed in some protoplasts after five days of culture. However, the binding of GFP-CBM27 to cell walls of divided cells and mature fronds were not clearly confirmed.

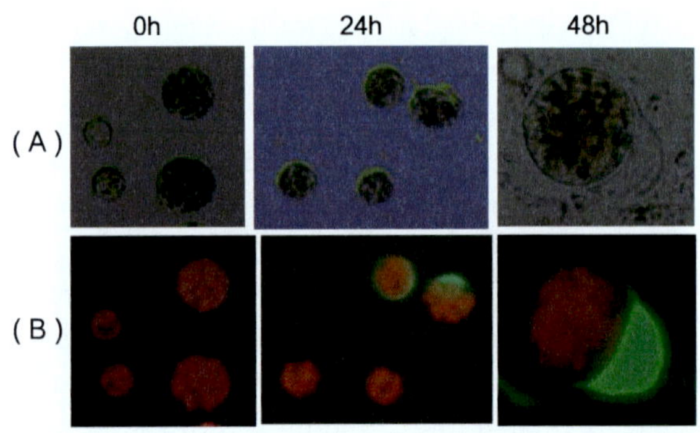

Figure 3. Binding of GFP-CBM27 to β-mannan in the regenerating cell walls of *B. atropurpurea* protoplasts. Protoplasts from different culture periods were used for GFP-CBM27 labeling. Immediately after protoplast isolation, they showed only the red autofluorescence of the photosynthetic pigments, and GFP-CBM27 linked to their regenerating cell walls emitted the green fluorescence.

To compare the binding patterns of the protoplasts with GFP-CBM27, CFW was also used to stain the regenerated cell walls of the protoplasts under the same condition. CFW did not bind to the cell walls within 24 h of culture. After 36 h of culture, CFW began to bind to the regenerated cell walls. However, the fluorescence of CFW on the cell surface was very faint and unclear as compared with that of GFP-CBM27 in the first 60 h of culture. These results indicate that GFP-CBM27 is more sensitive than CFW in the detection of amorphous β-mannan in the cell walls of *B. atropurpurea*. However, CFW bound to the cell walls of divided cells and mature fronds, whereas GFP-CBM27 did not bind to them (Umemoto and Araki 2010).

The cell walls of both *Bangia* and *Porphyra*, which belong to Bangiaceae, are composed of three kinds of unique polysaccharides: β-1,4-mannan, β-1,3-xylan, and porphyran (McDowell, 1967; Gretz et al., 1982). The cell wall polysaccharides of red algae play an important role not only in the structure but also in the defense system against several pathogens (Bouarab et al., 1999; Potin et al., 1999; Weinberger et al., 2000; Uppalapati and Fujita, 2000; Vreeland and Kloareg, 2000). Therefore, understanding the cell wall structure of these red algae, especially that of *Porphyra*, is essential for breeding and developing disease-resistant strains for cultivation. Although there have been some studies reporting the chemical composition of the cell wall polysaccharides of these red algae, very little is known about their complex structure and assembly in nature.

Recently, several non-catalytic CBMs have been reported to show great potential as specific and versatile probes of carbohydrates in the cell walls of higher plants (Hildén et al., 2003; McCartney et al., 2004; Daniel et al., 2006; McCartney et al., 2006; Filonova et al., 2007a, b). To date, however, there have been no studies analyzing algal cell walls by employing CBMs as probes. Therefore, in this study, we attempted to analyze the formation of cell walls of the red alga *B. atropurpurea* by using a GFP-CBM27 fusion protein in order to obtain insight into the cell wall structure of *Porphyra*. A mannan-binding CBM27 of Man5C from *Vibrio* sp. strain MA-138 was used for probing the distribution and assembly of β-mannan in the cell walls of *B. atropurpurea*. CBM27 has a specific ability to bind to soluble amorphous β-mannans but not to crystalline β-mannan (Tanaka et al., 2009). After the isolation of *B. atropurpurea* protoplasts, we labeled the protoplasts with the GFP-CBM27 fusion protein during their regeneration process and visualized the synthesis of β-mannan. Since protoplasts lack cell walls, they are very useful for investigating the process of cell wall regeneration, which is an important process for the subsequent development of the cell because cell division usually occurs after cell wall regeneration in protoplasts.

In a few protoplasts, the occurrence of β-mannan began to be observed at certain areas of their cell surface after 12 h of culture. As the protoplast culture progressed, β-mannans were spread on the entire surfaces of the cells. On the third day of cultivation, a majority of regenerating protoplasts showed fluorescence on their entire surfaces, while some protoplasts exhibited fluorescence strongly in the rhizoid region. It should be considered that protoplasts did not exhibit synchronous regeneration because protoplasts of different stages of growth were present. The ratio of protoplasts emitting fluorescence on their cell surfaces reached a plateau at approximately 30% after 48 h of culture in cells that were cultured in test tubes. On the other hand, when the protoplasts were cultivated on the cover glasses for 72 h, almost all of them formed cell walls. This could be because the protoplasts adhere to the internal surface of the test tubes through their regenerated cell walls; therefore, the ratio of protoplasts that generated cell walls did not increase above 30% in the culture fluid. It appears that the attachment of protoplasts to a substratum triggers cell wall formation and subsequent development.

CFW, which interacts with various polysaccharides (Wood, 1980), bound to the cell walls of divided cells and mature fronds as well as regenerated initial cell walls of the protoplasts, while the bindings of GFP-CBM27 to divided cells and mature fronds were not confirmed definitely. This has been considered that the crystal level of β-mannans in the cell wall becomes higher as the protoplast culture was continued, and that is why the CBM27 could not bind to divided cells and mature fronds, because CBM27 has a specific binding ability to soluble amorphous β-mannan but not to crystalline mannan. As the other reason, it has been also presumed that β-mannans in cell wall might be covered by other polysaccharides after cell divisions, because CFW has binding abilities to various polysaccharides such as crystalline β-mannan, β-1,3-xylan, and agarose as well as cellulose and chitin.

To date, four CBM families (families 23, 27, 29, and 35) that bind to soluble amorphous β-mannan have been identified (Stoll et al., 2000; Freelove et al., 2001; Sunna et al., 2001; Bolam et al., 2004). It has been reported that xylan-binding CBMs classified into the different CBM families showed differential recognition of plant cell walls (McCartney et al., 2006). Thus, if we used mannan-binding modules belong to other CBM families instead of CBM27, differential binding patterns of the regenerated cell walls may have been observed. Although CFW is a convenient and rapid stain for detecting plant cell walls (Nagata and Takebe, 1970; Galbraith, 1981; Hahne et al., 1983), it does not have the ability to distinguish between the different polysaccharides therein. On the other hand, CBMs have specific binding abilities to individual polysaccharides. Therefore, CBMs are excellent tools for understanding how an individual polysaccharide is formed during the regeneration process of algal protoplasts.

4. Conclusion

Porphyra protoplasts are useful tools for breeding by the cell fusion and the gene manipulation and for physiologic fundamental research. Recently, the protoplasts also attracted attention as material of a health food and the supplements of the cultured fish feed. The cell wall of *Porphyra* is composed of three kinds of polysaccharides (β-1,4-mannan, β-1,3-xylan, and porphyran). Therefore, the enzymes capable of degrading these polysaccharides are required for protoplast isolation from the *Porphyra* thalli. Although the digestive juices of marine Herbivores are often used for protoplast isolation, the activities of the polysaccharides-degrading enzymes in them are not so strong, and unevenness causes enzyme activity among their lots. Therefore, they have a fault that the efficiency for protoplast isolation is not good. In contrast, β-1,4-mannanase, β-1,3-xylanase, and agarase capable of degrading porphyran prepared from microorganisms are convenient for protoplast isolation, because the rate of each enzyme can be blended freely. It was found that the protoplast yield increased more than three times by preprocessing the thalli of *Porphyra* with protease before they were treated with the polysaccharide degrading enzymes. When the thalli of *Porphyra* were pretreated with the papain solution containing 5% papain and 0.7M mannitol at 22°C for 30 min and then treated with the cell wall degrading enzymes (five units of β-1,4-mannanase, two units of β-1,3-xylanase, two units of agarase, and 0.7M mannitol in 20mM MES buffer, pH 7.5) at 22°C for 90 min, 10^7order of protoplasts was isolated from the thalli of 1g wet weight. The C-terminal region that exists in β-mannanase-C of *Vibrio* sp.MA-138 was confirmed to be a CBM, which binds to β-1,4-mannan. The fusion protein of the CBM and GFP being a fluorescent dye was recognized to be a superior tool to observe reproduction of β-mannan on the protoplasts.

REFERENCES

Akino T, Nakamura N, and Horikoshi K (1988) Characterization of three β-mannanases of an alkalophilic *Bacillus* sp. *Agric. Biol. Chem.,* 52:773-779.

Aoki T, Araki T, and Kitamikado M (1988) Purification and characterization of an endo-β-1,3-xylanase from *Vibrio* sp. *Nippon suisan gakkaishi.* 54:277-281.

Aoki T, Araki T, and Kitamikado M (1990a) Purification and characterization of β-agarase from *Vibrio* sp. AP-2. *Nippon Suisan Gakkaishi.,* 56:825-830.

Aoki T, Araki T and Kitamikado M (1990b) Purification and characterization of a novel β-agarase from *Vibrio* sp. Ap-2. *Eur. J. Biochem*., 187: 461-465.

Aoki Y and Kamei Y (2006) Preparation of recombinant polysaccharide-degrading enzymes from the marine bacterium, *Pseudomonas* sp. ND137 for the production of protoplasts of *Porphyra yezoensis*. *Eur. J. Phycol.*, 41: 321–328.

Araki T and Kitamikado M (1978) Distribution of mannan-degrading bacteria in aquatic environments, *Bull. Japan. Soc. Sci. Fish.*, 44: 1135-1139.

Araki T, Aoki T, Kitamikado M (1987) Preparation and regeneration of protoplasts from wild-type of *Porphyra yezoensis* and green variant of *P. tenera. Nippon Suisan Gakkaishi*, 53: 1623–1627.

Araki T and Morishita T (1990) Fusion of protoplast from wild type *Porphyra yezoensis* and green type *P. tenera* thalli (Rhodophyta). *Nippon Suisan Gakkaishi*, 56: 1161.

Araki T, Hayakawa M, Tamaru T, Yoshimatsu K and Morishita T (1994) Isolation and regeneration of haploid protoplasts from *Bangia atropurpurea* (*Rhodophyta*) with marine bacterial enzymes. *J. Phycol.,* 30:1040-1046.

Araki T, Inoue N, and Morishita T (1998a) Purification and characterization of β-1,3-xylanasa from a marine bacterium, *Alcaligenes* sp. XY-234. *J. Gen. Appl. Microbiol.,* 44:269-274.

Araki, T., M. Hayakawa, L. Zhang, S. Karita, and T. Morishita (1998b) Purification and characterization of agarases from a marine bacterium, *Vibrio* sp. PO-303. *J. Mar. Biotechnol.,* 6:260-265.

Araki T, Tani S, Maeda K, Hashikawa S, Nakagawa H, and Morishita T (1999) Purification and characterization of β-1,3-xylanase from a marine bacterium, *Vibrio* sp. XY-214 , *Biosci. Biotechnol. Biochem.*, 63: 2017-2019.

Araki T, Hashikawa S, and Morishita T (2000) Cloning, sequencing, and expression in *Escherichia coli* of the new gene encoding β-1,3-xylanase from a marine bacterium, *Vibrio* sp. strain XY-214, *Appl. Environ. Microbiol.,* 66: 1741-1743.

Aspinall GO, Hirst EL, Percival EGV, and Williamson IR (1953) The mannans of ivory nut (*Phytelephas macrocarpa*) part I. The methylation of mannan A and mannan B. *J. Chem. Soc.*, 3184-3188.

Bastawde KB (1992) Xylan structure, microbial xylanases, and their mode of action. *World Journal of Microbiology and Biotechnology*, 8: 353-368.

Baumann P, Furniss AL, and Lee JV (1984) pp.518-538 in "Bergey's Manual of Systematic Bacteriology" (ed. by Krig and Holt JG) Vol. 1, Williams and Wilkins, U.S.A.

Edward A. Bayer EA, Belaich JP, Shoham Y, and Lamed R (2004) THE CELLULOSOMES: Multienzyme Machines for Degradation of Plant Cell Wall Polysaccharides. *Annual Review of Microbiology*, 58: 521-554.

Beg QK, Kapoor M, Mahajan L, Hoondal GS (2001) Microbial xylanases and their industrial applications: a review. *Applied Microbiology and Biotechnology*, 56: 326-338.

Belas, R. (1989) Sequence analysis of the *agrA* gene encoding beta-agarase from *Pseudomonas atlantica*. *J. Bacteriol,* 171: 602-605.

Bibb MJ, Jones GH, Joseph R, Buttner MJ and Ward JM (1987) The agarase gene (*dagA*) of *Streptomyces coelicolor* A3(2): affinity purification and characterization of the cloned gene products. *J. Gen. Microbiol.*, 133:2089-2096.

Bolam DN, Xie HF, Pell G, Hogg D, Galbraith G, Henrissat B, and Gilbert HJ (2004) X4 modules represent a new family of carbohydrate-binding modules that display novel properties. *J. Biol. Chem.*, 279:22953-22963.

Bouarab K, Potin P, Correa J, and Kloareg B (1999) Sulfated oligosaccharides mediate the interaction between a marine red alga and its green algal pathogenic endophyte. *The Plant Cell,* 11:1-17.

Chawla HS (2002) Protoplast Isolation, p.84, In "Introduction to plant biotechnology" Science Publishers, Inc. New Hampshire.

Chen WP, Matsuo M, and Yasui T (1986) Purification and some properties of □-1,3-xylanase from *Aspergillus terreus* A-07. *Agric. Biol. Chem.*, 50:1183-1194.

Cocking EC (1966) An electron microscopic study of the initial stages of infection of isolated tomato fruit protoplasts by tobacco mosaic virus. *Planta.*, 68: 206-214.

Coutinho PM and Henrissat B (1999) The modular structure of cellulases and other carbohydrate-active enzyme: an integrated database approach, p.15-23. In Ohmiya K, ayashi K, Sakka K, Kobayashi C, Karita S, Kimura T, Genetics, biochemistry and ecology of cellulose degradation, Uni Publishers Co., Tokyo, Japan.

Craigie, J (1990) Chapter 10 Cell Walls, p.221-258, In S. R (ed.), Biology of the Red Algae. Cambridge University Press, New York.

Daniel G, Filonova L, Kallas AM, and Teeri TT (2006) Morphological and chemical characterization of the G-layer in tension wood fibers of *Populus tremula* and *Betula verrucosa*: Labeling with cellulose-binding module CBM 1*HjCel7A* and fluorescence and FE-SEM microscopy. *Holzforchung,* 60:618-624.

Doi RH and Kosugi A (2004) Cellulosomes: plant-cell-wall-degrading enzyme complexes. *Nature Reviews Microbiology,* 2: 541-551.

Doi RH (2008) *Cellulases of Mesophilic Microorganisms Cellulosome and Noncellulosome Producers.* Ann. N.Y. Acad. Sci. 1125: 267–279.

Dong J, Hashikawa S, Konishi T, Tamaru Y and Araki T (2006a) Cloning of the novel gene encoding β-agarase C from a marine bacterium, *Vibrio* sp. strain PO-303, and characterization of the gene product. *Appl. Environ. Microbiol.,* 72:6399-6401.

Dong J, Tamaru Y, and Araki T (2006b) Molecular cloning, expression, and characterization of a β–agarase gene, *agaD*, from a marine bacterium, *Vibrio* sp. strain PO-303. *Biosci Biotechnol Biochem,* 71, 38-46.

Dong J, Tamaru Y and Araki T (2007) A unique β-agarase, AgaA, from a marine bacterium, *Vibrio* sp. PO-303. *Appl Microbiol Biotechnol,* 74: 1248-1255.

Drew KM (1949) Conchocelis-phase in the life-history of Porphyra umbilicalis (L.) Kütz. *Nature,* 164: 748-749.

Duffaud GD, Mccutchen CM, Leduc P, Parker KN, and R. M. Kelly RM (1997) Purification and characterization of extremely thermostable beta-mannanase, beta-mannosidase, and alpha-galactosidase from the hyperthermophilic eubacterium *Thermotoga neapolitana* 5068. *Appl. Environ. Microbiol.,* 63:169-177.

Emi S, Fukumoto J, and Yamamoto T (1972) Crystallization and some properties of mannanase. *Agric. Biol. Chem.,* 36: 991-1001.

Filonova L, Kallas AM, Greffe L, Johansson G, Teeri TT, and Daniel G (2007b) Analysis of the surfaces of wood tissues and pulp fibers using carbohydrate-binding modules specific for crystalline cellulose and mannan. *Biomacromolecules,* 8: 91-97.

Freelove AC, Bolam DN, White P, Hazlewood GP, and Gilbert HJ (2001) A novel carbohydrate-binding protein is a component of the plant cell wall-degrading complex of *Piromyces equi. J. Biol. Chem.,* **276**: 43010-43017.

Frei E and Preston RD (I964) Non-cellulosic structural polysaccharides in algal cell walls. II. Association of xylan and mannan in *Porphyra umbilicalis. Proc. R. Soc.* Lond. B, 160: 293-313.

Fujita Y and Migita S (1985) Isolation and culture of protoplasts from some seaweeds, *Bull. Fac. Fish.* Nagasaki Univ. 57: 39-45.

Fujita Y and Saito M (1990) Protoplast isolation and fusion in *Porphyra* (Bangiales, Phodophyta), *Hydrobiologia,* 204/205: 161-166.

Galbraith DW (1981) Microfluorimetric quantitation of cellulose biosynthesis by plant protoplasts using Calcofluor White. *Physiol. Plant.,* 53: 111-116.

Gall EA, Chiang YM, and Kloareg B (1993) Isolation and regeneration of protoplasts from *Porphyra dentate* and *Porphyra crispata. Eur.. J. Phycol.,* 28: 277-283.

Gretz MR, Sommerfeld MR, and Aronson JM (1982) Cell wall composition of the genetic phase of *Bangia atropurpurea* (Rhodophyta). *Bot. Mar.,* 25: 529-535

Hahne G, Herth W, and Hoffmann F (1983) Wall formation and cell division in fluorescence-labeled plant protoplasts. *Protoplasma,* 115: 217-221.

Hayashi H, Takagi K, Fukumura M, Kimura T, Karita S, Sakka K, and Ohmiya K (1997) Sequence of xynC and properties of XynC, a major component of the *Clostridium thermocellum* cellulosome. *J. Bacteriol.,* 179: 4246-4253.

Henrissat B (1991) A classification of glycosyl hydrolases based on amino acid sequence similarities. *Biochem. J.,* 280: 309-316.

Henrissat B and Bairoch A (1996) Updating the sequence-based classification of glycosyl hydrolases. *Biochem. J.,* 316: 695-696.

Hildén L, Daniel G, and Johansson G (2003) Use of a fluorescence labeled, carbohydrate-binding module from *Phanerochaete chrysosporium* Cel7D for studying wood cell wall ultrastructure. *Biotechnol. Lett.,* 25: 553-558

Hosoda A, Sakai M, and Kanazawa S (2003) Isolation and characterization of agar-degrading *Paenibacillus* spp. associated with the rhizosphere of spinach. *Biosci. Biotechnol. Biochem.,* 67: 1048-1055.

Iriki Y, Suzuki T, Nisizawa K and Miwa T (1960) Xylan od Siphonaceous green algae. *Nature*, 187: 82-83.

Jam M, Flament D, Allouch J, Potin P, Thion L, Kloareg B, Czjzek M, Helbert W, Michel G, and Barbeyron T (2005) The endo-□-agarases AgaA and AgaB from the marine bacterium *Zobellia galactanivorans*: two paralogue enzymes with different molecular organization and catalytic behaviors. *Biochem. J.,* 385:703-713.

Jones JKN (1950) The structure of the mannan present in *Porphyra umbilicalis. J. Chem. Soc.,* 3292-3295.

Kalla A, Yoshimatsu T, Araki T, Dong MZ, Yamamoto T, Tuyoshi Sugita T, and Sakamoto S (2008a) Utilization of *Porphyra* spheroplasts as a feed additive for red sea bream, *Fisheries Science*, 74: 104-108

Kalla A, Yoshimatsu T, Higano J, Umemoto Y, Sakamoto S and Araki T (2008b) Novel Diet for Short-Neck Clam by Using *Porphyra* Spheroplasts, *Aquaculture Sci.*, 56: 465-466.

Khan MND, Yoshimatsu T, Kalla A, Araki T and Sakamoto S (2008) Supplemental effect of *Porphyra* spheroplasts on the growth and feed utilization of black sea bream. *Fisheries Science*, 74, 397-404.

Love J and Percival E (1964) The polysaccharides of green seaweed *Codium fragile*. Part III. A β-linked mannan. *J. Chem. Soc.,* 3345-3350.

Maeda H and Ishida N (1967) Specificity of binding of hexapyranosyl polysaccharides with fluorescent brightener. *J. Biochem.,* 62: 276-278.

McCartney L, Gilbert HJ, Bolam DN, Boraston AB, and Knox JP (2004) Glycoside hydrolase carbohydrate-binding modules as molecular probes for the analysis of plant cell wall polymers. *Anal. Biochem.,* 326: 49-54

McCartney L, Blake AW, Flint J, Bolam DN, Boraston AB, Gilbert HJ, and Knox JP (2006) Differential recognition of plant cell walls by microbial xylan-specific carbohydrate-binding modules. *Proc. Natl. Acad. Sci.,* USA 103: 4765-4770.

McDowell RH (1967) Chemistry and Enzymology of Marine Algal Polysaccharides. Academic Press, London, pp.88-96, 134-137.

Van der Meulen HJ and Harder W (1975) Production and characterization of the agarase of *Cytoplaga flevensis*. *Antonie Van Leeuwenhoek.* 41: 431-437.

Minocha SC (2001) Genetic engineering of seaweeds: current status and perspectives, In Proceedings of the 17th International Seaweed Symposium, Anthony R. O. Chaoman, p. 19-29, Oxford University Press, Ingrand.

Mizukami Y, Okauchi M and Kito H (1992) Effects of cell wall-lytic enzymes on the electrofusion efficiency of protoplasts from *Porphyra yezoensis*. *Aquaculture*, 108, 193-205.

Morrice LM, McLean MW, Long WF and Williamson FB (1983a) Beta-agarses I and II from *Pseudomonas atlantica*. Purifications and some properties. *Eur. J. Biochem.*, 135: 553-558.

Morrice LM, McLean MW, Long WF and Williamson FB (1983b) Beta-agarases I and II from *Pseudomonas atlantica*. Substrate specificities. *Eur. J. Biochem.,* 137: 149-154.

Nagata T, Takebe I (1970) Cell wall regeneration and cell division in isolated tobacco mesophyll protoplasts. *Planta,* 92: 301-308.

Oda Y, Komaki T and Tonomura K (1993) Purification and properties of extracellular β-mannanases produced by *Enterococcus casseliflavus* FL2121 isolated from decayed konjac. *J. Ferment. Bioengin.*, 76: 14-18.

Ohta Y, Hatada Y, Nogi Y, Miyazaki M, Li Z, Akita M, Hidaka Y, Goda S, Ito S and Horikoshi K (2004b) Enzymatic properties and nucleotide and amino acid sequence of a thermostable β-agarase from a novel species of deep-sea *Microbulbifer*. *Appl. Microbiol. Biotechnol.*, 64: 505-514.

Okazaki F, Tamaru Y, Hashikawa S, Li YT and Araki T (2002) Novel Carbohydrate-Binding Module of □-1,3-Xylanase from a Marine Bacterium, *Alcaligenes* sp. Strain XY-234. *Journal of Bacteriology,* 184: 9, 2399–2403

Percival E and McDowell RH (1967) Chemistry and enzymology of marine algal polysaccharides. Academic Press, London. pp. 88-98, 129-137.

Polne-Fuller M and Gibor A (1984) Developmental studies in *Porphya perforata* as expressed by morphology, enzymatic digestion, and protoplast regeneration. *J. Physiol,* 20: 609-616.

Potin P, Bouarab K, Küpper F and Kloareg B (1999) Oligosaccharide recognition signals and defense reactions in marine plant-microbe interactions. *Curr Opin Microbiol.,* 2: 276-283

Provasoli L, McLaughlin JJA and Droop MR (1957) The development of artificial media for marine algae. *Arch. Mikrobiol.*, 25: 392-428.

Punja ZK (2001) Genetic engineering of plants to enhance resistance to fungal pathogens— a review of progress and future prospects. *Canadian Journal of Plant Pathology*, 23: 216 – 235.

Rees DA (1961) Enzymic Synthesis of 3:6-Anhydro-L-Galactose within Porphyran from L-Galactose 6-Sulphate Units. *Biochem. J.,* 81: 347.

Saga N (1984) Isolation of protoplasts from edible seaweeds, *Bot. Mag. Tokyo,* 97: 423-427,

Saga N and Sakai Y (1984) Isolation of protoplasts from *Laminaria* and *Porphyra*. *Bull. Jpn. Soc. Sci. Fish.,* 50: 1085.

Sateesh MK (2003) Somatic Hybridization and Endospwem Culture p179 In Biotechnology-5: Animal Cells, Immunology and Plant Biotechnology. New Age International.

Scheller HV and Ulvskov P (2010) Hemicelluloses. *Annual Review of Plant Biology,* 61: 263-289.

Schroeder DC, Jaffer MA and Coyne VE (2003) Investigation of the role of a beta(1-4) agarase produced by *Pseudoalteromonas gracilis* B9 in eliciting disease symptoms in the red alga *Gracilaria gracilis*. *Microbiology,* 149: 2919-2929.

Shieh WY and Jean WD (1998) *Alterococus agarolyticus*, gen.nov., sp.nov., a halophilic thermophilic bacterium capable of agar degradation. *Can. J. Microbiol.,* 44: 637-645.

Shine J and Dalgarno L (1975) Determinant of cistron specificity in bacterial ribosomes. *Nature,* 254:34-38.

Somogyi M (1952) Notes on sugar determination. *J. Biol. Chem.,* 195: 19-23.

Stoll DB, Stalbrand H, McLean BW, Kilburn DG and Warren RAJ (2000) Mannanase Man26A from *Cellulomonas fimi* has a mannan-binding module. *FEMS Microbiol Letters,* **183**:265-269.

Su J and Hassid WZ (1962) Carbohydrates and nucleotides in the red alga, *Porphyra perforate*. I. Isolation and identification of carbohydrates. *Biochemistry,* 1: 468-474.

Sugano Y, Matsumoto T, Kodama H and M. Noma M (1993a) Cloning and sequencing of *agaA*, a unique agarase 0107 gene from a marine bacterium, *Vibrio* sp. strain JT0107. *Appl. Environ. Microbiol.*, 59: 3750-3756.

Sugano Y, Terada I, Arita M, Noma M and Matsumoto T (1993b) Purification and characterization of a new agarase from a marine bacterium, *Vibrio* sp. strain JI0107. *Appl Environ Microbiol.*, 59: 1549-1554.

Sunna A, Gibbs MD and Bergquist PL (2001) Identification of novel beta-mannan- and beta-glucan-binding modules: evidence for a superfamily of carbohydrate-binding modules. *Biochem. J.*, 356: 791-798.

Takahashi R, Kusakabe I, Kobayashi H, Murakami K, Maekawa A, and Suzuki T (1984) Purification and some properties of mannanase from *Streptomyces* sp. *Agric. Biol. Chem.*, 48: 2189-2195.

Takebe I, Otsuki Y and Aoki S (1968) Isolation of tobacco mesophyll cells in intact and active state, *Plant Cell Physiol.*, 9: 115-124.

Talbot G and Sygusch J (1990) Purification and characterization of thermostable beta-mannanase and alpha-galactosidase from *Bacillus stearothermophilus*. *Appl. Environ. Microbiol.*, 56: 3505-3510.

Tamaru Y, Araki T, Amagoi H, Mori H and Morishita T (1995) Purification and characterization of an extracellular beta-1,4-mannanase from a marine bacterium, *Vibrio* sp. strain MA-138. *Appl. Environ. Microbiol.*, 61: 4454-4458.

Tamaru Y, Araki T, Morishita T, Kimura T, Sakka K, and Ohmiya K (1997) Cloning, DNA sequencing, and expression of a □-1,4-mannanase gene from a marine bacterium, *Vibrio* sp. strain MA-138. *J. Ferment. Bioeng.*, 83: 201-205.

Tao C, Sugawara T, Maeda S, Wang X and Hirata T (2008) Antioxidative activities of a mycosporine-like amino acid, porphyra-334. *Fisheries Science* 74: 1166-1172.

Tanaka M, Umemoto Y, Okamura H, Nakano D, Tamaru Y and Araki T (2009) Cloning and Characterization of a β-1,4-Mannanase 5C Possessing a Family 27 Carbohydrate-Binding Module from a Marine Bacterium, *Vibrio* sp. Strain MA-138. *Biosci. Biotechnol. Biochem.*, 73: 109-116.

Umemoto Y and Araki T (2010) Cell Wall Regeneration in *Bangia atropurpurea* (Rhodophyta) Protoplasts Observed Using a Mannan-Specific Carbohydrate-Binding Module. *Marine Biotechnology,* 12, 24-31.

Uppalapati SR and Fujita Y (2000) Carbohydrate regulation of attachment, encystment, and appressorium formation by *Pythium porphyrae* (Oomycota) zoospores on *Porphyra yezoensis* (Rhodophyta). *J. Phycol.*, 36:359-366.

Usov AI and Miroshnikova LI (1975) Isolation of agarase from *Littorina mandshurica* by affinity chromatography on Biogel A. *Carbohydr. Res.*, 43:204-207.

Vera J, Alvarez R, Murano E, Slebe JC and Leon O (1998) Identification of a marine agarolytic *Pseudoalteromonas* isolate and characterization of its extracellular agarase. *Appl. Environ. Microbiol.*, 64: 4378-4383.

Villarroya H and S Petek F (1976) Purification and properties of a β-mannanase from alfalfa seeds. *Biochimica et Biophysica Acta.*, 7: 200-211

Vreeland V and Kloareg B (2000) Cell wall biology in red algae: divide and conquer. *J. Phycol.*, 36: 793-797

Weinberger F and Friedlander M (2000) Response of *Gracilaria conferta* (Rhodophyta) to oligoagars results in defense against agar-degrading epiphytes. *J. Phycol.,* 36:1079-1086.

Wolfrom ML, Laver ML and Patin DL (1961) Carbohydrate of the coffee beans II. Isolation and characterization of a mannan. *J. Org. Chem.*, 26: 4533-4535.

Wong KK, Tan LU and Saddler JN (1998) Multiplicity of β-1,4-xylanase in microorganisms: functions and applications. *Microbiology and Molecular Biology Reviews,* 52: 305–317.

Wood PJ (1980) Specificity in the interaction of direct dyes with polysaccharides. *Carbohydr. Res.,* 85: 271-287

Yamaguchi K, Araki T, Aoki T, Tseng CH, and Kitamikado M (1989) Agal cell wall-degrading enzymes from viscera of marine animals. *Noppon Suisan Gakkaishi,* 55: 105-110.

Yamaura I, Matsumoto T, Funatsu M and Funatsu Y (1990) Purification and some properties of endo-1,4-β-D-mannanase from *Pseudomonas* sp. PT-5. *Agric. Biol. Chem.,* 54: 2425-2427.

Yoshizawa Y, Ametani A, Tsunehiro J, Nomura K, Itoh M, Fukui F and Kaminogawa S. (1995) Macrophage stimulation activity of the polysaccharide fraction from a marine alga (*Porphyra yezoensis*): structure-function relationships and improve solubility. *Biosci. Biotechnol. Biochem.,* 59:1933-1937.

In: Porphyra yezoensis
Editor: Koji Mikami

ISBN 978-1-61122-304-0

Chapter 9

BIOTECHNOLOGICAL APPLICATION OF PORPHYRA YEZOENSIS GENES

Tadatake Oku*[*]*, Hirotaka Chida, Hideharu Akazaki, Shin-ichiro Yamamoto, Kazufusa Shinomiya*[1]*, and Hidetoshi Sakurai

Graduate School of Bioresource Sciences, Nihon University,
1866, Kameino, Fujisawa-shi, Kanagawa 252-0880, Japan
[1]Graduate School of Pharmacy, Nihon University,
7-7-1, Narashinodai, Funabashi-shi, Chiba 274-8555, Japan

ABSTRACT

An approaching of the fundamental and applied aspects of photosynthesis is now indispensable to a global range of science and technology. Photosynthesis occurs in two phases, the light reactions and the dark reactions. Photosynthetic plants convert light energy into ATP and NADPH in the photosynthetic electron transfer and photophosphorylation, and synthesize mainly carbohydrates in the Calvin-Benson cycle. In this chapter, we show the enhancement of photosynthesis and growth of plants by introducing the gene of a red algal *Porphyla yezoensis* cytochrome c_6, which has been evolutionally eliminated from higher plant chloroplasts, into the model plant *Arabidopsis thaliana*. At 60 days after planting, the plant height, leaf and root lengths of the transformants were 1.3-, 1.1- and 1.3-fold those in the wild type plants, respectively. At the same time, in the transgenic plants, the amounts of chlorophyll, protein, ATP, NADPH and starch were 1.2-, 1.1-, 1.9-, 1.4-, and 1.2-fold those in the wild type plants, respectively. The CO_2 assimilation capacity of the transgenic plants was 1.3-fold than that of the wild-type. Moreover, in the transgenic *Arabidopsis* expressing algal cytochrome c_6, the 1-qP, which reflects the reduced state of plastoquinone pool, there is 30% decrease than that in the wild-type. These results show that the electron transfer of photosynthesis of *Arabidopsis* would be accelerated by the expression of algal

[*] Correspondence to: Tadatake Oku, Graduate School of Bioresource Sciences, Nihon University, 1866 Kameino, Fijisawa-shi, Kanagawa 252-0880, Japan Tel/Fax: +81-466-84-3950, E-mail: oku@brs.nihon-u.ac.jp, okutad@ jcom.home.ne.jp.

cytochrome c_6. Our results demonstrate that the growth and photosynthesis of *Arabidopsis* plants could be enhanced by the expression of algal cytochrome c_6 gene. We also show interesting brown algal and diatom cytochrome c_6 beside a red algal *Porphyla yezoensis* cytochrome c_6.

ABBREVIATIONS

Cyt – cytochrome;
ATP- adenosine 5'-triphosphate;
NADPH- nicotinamide adenine dinucleotidephosphate;
PS- photosystem; PC- plastocyanin;

I. INTRODUCTION

Photosynthesis takes place in green plants, algae and photosynthetic bacteria where the sunlight is converted into chemical energy. It is generally believed that the first photosynthetic organisms appeared being the photosynthetic bacteria and primitive algae about 3.5 billion years ago. Cyanobacteria evolved about two billion years ago, leading to the accumulation of oxygen in the atmosphere. All our food and fossil and fuels are derived from the process of photosynthesis. An increase in the concentration of carbon dioxide gas in atmosphere is causing a rise in global temperature and could bring about change in climate. Even modern industry is highly dependent on the products of photosynthesis because oil, coal, and gas are thought to be the remains of ancient organisms. Thus, the examination of the efficiency of the solar energy utilization by photosynthesis is now essential for development of science and technology such as biomass production, crop production, reduction and fixation of CO_2 and so forth. The site of photosynthesis in higher plant and algae is the chloroplast. Chloroplasts, of which there are 1～1,000 per cell, vary considerably in size and shape but are typically ～5 μ m long ellipsoids and have a highly permeable outer membrane and an inner membrane separated by a narrow intermembrane space. The inner membrane encloses the stroma. The light reactions, photosynthetic electron transport and phosphorylation occur in the thylakoid membranes forming the grana in Figure 1. Electron flow during the light-dependent stage of photosynthesis cells Z scheme in Figure 2. The electron flow system consists of three protein complexes: PSII, the cytochrome *b6f* complex, and PSI. The hemeprotein cytochrome (Cyt) c_6 and/or copper protein plastocyanin (PC) transfer electrons between Cyt *b6f* complex and photosystem I (PSI) reaction center complex in algal photosynthetic electron transport chain (Katoh, 1960a; Katoh and Takamiya, 1961). Although the structures of the two proteins are totally different, they are approximately the same in molecular weight and redox potential (Kerfeld and Krogman, 1998; Ullmann et al., 1997). Cyt c_6 (formerly named Cyt c_{553}) has been first prepared and crystallized from a red alga in 1960 (Katoh, 1960b). In some cyanobacteria and green algae, Cyt c_6 replaces PC in response to copper deficiency (Merchant and Bogorad, 1987). On the other hand, this does not occur in land plants, and it is widely accepted that Cyt c_6 has been evolutionally eliminated from land plants (De la Rosa et al., 2002; Sigfridsson, 1998). However, in 2002, genomic analysis and expression sequence tags (EST) analysis revealed a gene encoding Cyt c_6-like protein (Cyt c_{6A} or Cyt c_P) in higher plants (Wastl et al., 2002; Howe et al., 2006; De la Rosa et al., 2006). Although it was proposed that Cyt c_{6A} could substitute for PC in electron transfer in *Arabidopsis* (Gupta, 2002), electron transfer experiments and a genetic analysis of *Arabidopsis* mutants (Molina-Heredia et al., 2003; Weigel et al., 2003) are directly opposed to this view.

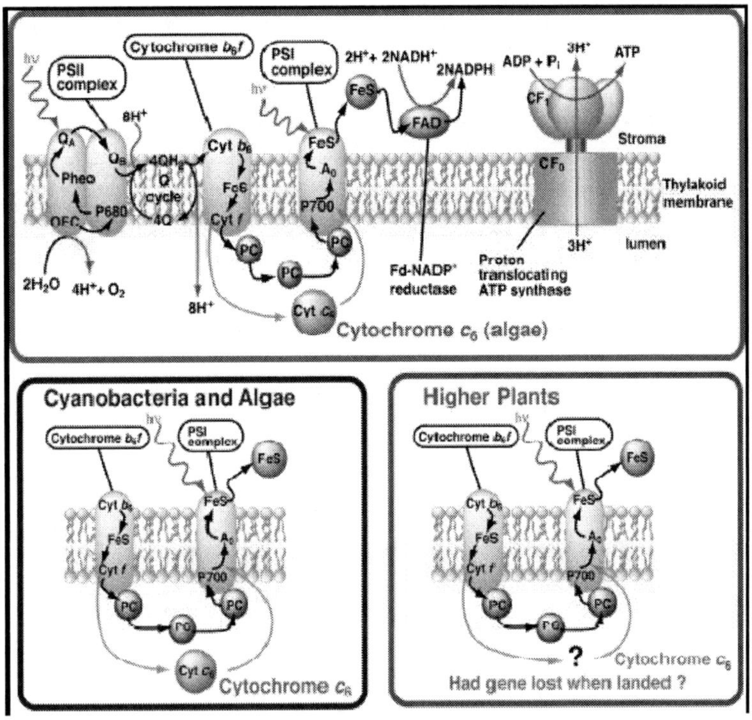

Figure 1. Scheme for Thylakoid membrane showing the components of its electron transport chain.

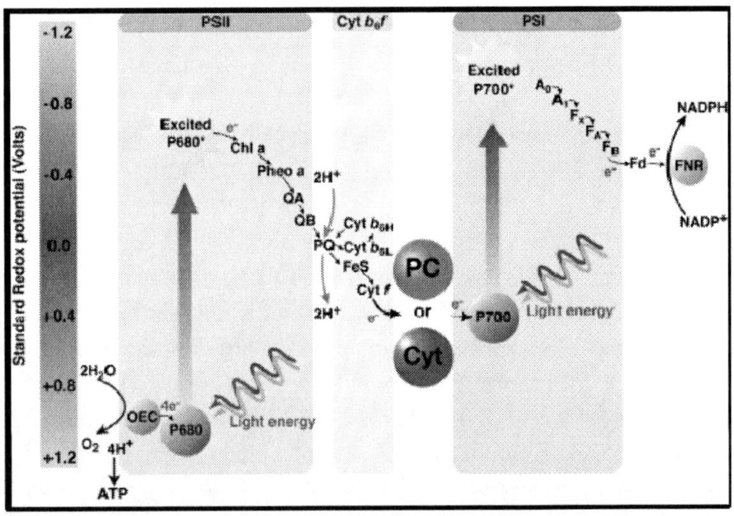

Figure 2. Z-scheme for photosynthetic electron transfer chain.

On the basis of the tertiary structure of Cyt c_{6A} and on its low redox potential, we also showed that the Cyt c_{6A} is an unsuitable electron donor between Cyt b_6f and PSI (Chida et al., 2006). We assume that PC is the only electron carrier that is effectively working *in vivo*.

There are many reports on introducing gene of Calvin-Benson cycle enzymes of photosynthesis into plants, but there are only a few studies that have been shown to enhance carbon dioxide (CO_2) fixation and growth in transgenic plants (Miyagawa et al., 2001; Ku et al., 1999). On the other hand, in the photosynthetic electron transfer and photophosphorylation of plants, there have been very few reports on the enhancement of photosynthesis and growth based on the increase in nicotinamide adenine dinucleotide phosphate (NADPH) by gene induction (Ramiro et al., 2007), but no reports on that of adenosine 5'-triphosphate (ATP).

We try to introduce algal Cyt c_6 gene to the model plant *Arabidopsis thaliana*, thus going back in the evolutional history of plants (Chida et al., 2007). Electron transfer of red algal Cyt c_6 and *Arabidopsis* PC to *Arabidopsis* PSI (Hiyama, 2004; Mackinney, 1941) is first examined *in vitro*, and we subsequently introduce the *Porphyra yezoensis* Cyt c_6 gene in higher plants *A. thaliana* (Theologis et al., 2000; Dandekar et al., 2005). We demonstrate that the expression of Cyt c_6 in addition to that of PC in chloroplasts lead to a high growth rate measured in terms of height, roots and leaf lengths, an increase in the amounts of chlorophylls, proteins, ATP, NADPH, starch, and increase in the capacity of CO_2 assimilation and the efficiency of the photosynthetic electron transfer. In this chapter, we present the reinforcement of photosynthetic electron transfer and photophosphorylation by introducing algal Cyt c_6 gene to *A. thaliana*. Moreover, in item V of this chapter, we reveal our recent data concerning physicochemical properties and crystal structures of Cyt c6 from the brown alga *Hizikia fusiformis* (Akazaki et al., 2008) and the diatom *Phaedactylum tricornutum* (Akazaki et al., 2009) in comparison to them of *P. yezoensis* (Yamada et al., 2000).

II. Electron Transfer of *Porphyra* Cyt *c6* to *Arabidopsis* PSI In Vitro

The Cyt c_6 from the red alga *P. yezoensis* was selected for *in vitro* experiments, since the midpoint redox potential of this cytochrome (+371 mV) (Yamada et al., 2000; Satoh et al., 2002) is similar to that of *A. thaliana* PC (+381 mV) (Molina-Heredia et al., 2003). Photooxidation of P700, the reaction center chlorophyll of PSI, can be monitored by the increase in the absorbance 820 nm upon illuminated by a Xe-flash. P700 could be re-reduced by Cyt c_6 or PC, which is the soluble electron donor to PSI, with concentration-dependent manner (Hervás et al., 1994). To investigate whether algal Cyt c_6 could substitute for the land plants PC in its physiological role, we have performed a flash-induced kinetic analysis of the *A. thaliana* PSI preparation with *P. yezoensis* Cyt c_6 and *A. thaliana* PC. The half-decay times of dark relaxation of the absorbance change due to P700 in the presence of 10 μM *P. yezoensis* Cyt c_6 were 4.4-fold shorter than *A. thaliana* PC (Figure 3). The difference between the half-decay times were 3.1, 3.0, 3.9, 3.9 and 4.4-fold at the concentration of 1, 2, 5, 15 and 20 μM electron donors, respectively. These results indicate that the *Porphyra* Cyt c_6 can effectively transfer an electron to *Arabidopsis* PSI.

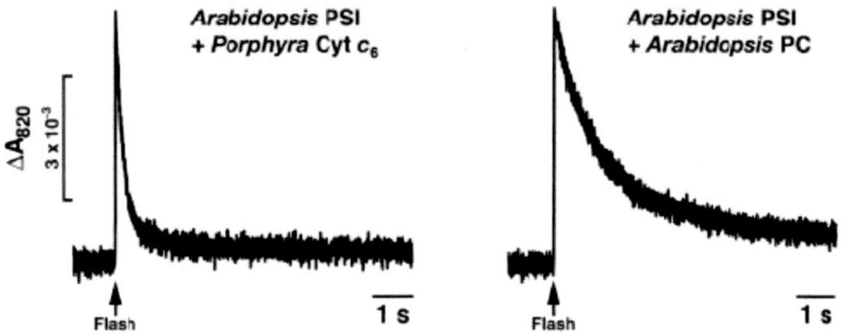

Figure 3. Kinetic trace of *Arabidopsis* PSI reduction by *Porphyra* cytochrome c_6 and *Arabidopsis* plastocyanin *in vitro*. The PSI particles were isolated from *A. thaliana*. Metalloprotein concentration was 10 μM.

In vitro experiments first demonstrate that Cyt c_6 from the red alga *P. yezoensis* can transfer an electron to *Arabidopsis* PSI, and the half life (time) of *Arabidopsis* PSI reduction of *Porphyra* Cyt c_6 was shorter than that of *Arabidopsis* PC (Figure 3). Likewise, in the cyanobacteria *Pseudanabaena* sp. PCC6903, the half life (time) of *Pseudanabaena* PSI of Cyt c_6 was shorter than that of PC (Hervás et al., 1998). It has been suggested that release of oxidized PC from PSI limits electron transfer between Cyt b_6f and PSI (Finazzi et al., 2005). The shorter half life (time) of the dark reduction kinetics of *Arabidopsis* PSI by *Porphyra* Cyt c_6 compared with *Arabidopsis* PC (Figure 3) could be attributed to be faster release of oxidized *Porphyra* Cyt c_6 from *Arabidopsis* PSI compared with *Arabidopsis* PC, since our excitation flash is not single-turnover for the PSI reaction center.

III. Development of Transgenic *Arabidopsis* with Cyt *c6* and Acceleration of Growth and Amount of Metabilotes of Plants

We produced transgenic *Arabidopsis* plant expressing *Porphyra* Cyt c_6 gene, which reduces *Arabidopsis* PSI more efficiently compared with *Arabidopsis* PC *in vitro* experiments as described above. To target Cyt c_6 into the chloroplast lumen, we constructed a plasmid containing the Cyt c_6 gene from the red alga *P. yezoensis* with the *petE* (plastocyanin) transit peptides (Hageman et al., 1986; Soll and Schleiff, 2004) from *A. thaliana*. The constructed plasmid DNA was introduced into the *A. thaliana* plants by *Agrobacterium* method (Dandekar and Fisk, 2005). We next introduce the *Porphyra* Cyt c_6 into the model plant *A. thaliana* (Figs. 4a, 4b 4c and 4d). This is the first report on the expression of *Porphyra* Cyt c_6 in *A. thaliana* chloroplasts. Then, we found that reinforcement of photosynthetic electron transfer and photophosphorylation is caused by introducing *Porphyra* Cyt c_6 gene into the model plant *A. thaliana*. Genomic DNA from the several transformed plants was isolated and analyzed by Polymerase Chain Reaction (PCR) to check if the Cyt c_6 gene was inserted into the genome (Figure 4b). Approximately 0.5% of the analyzed primary transgenic plants (T1) had the insertion.

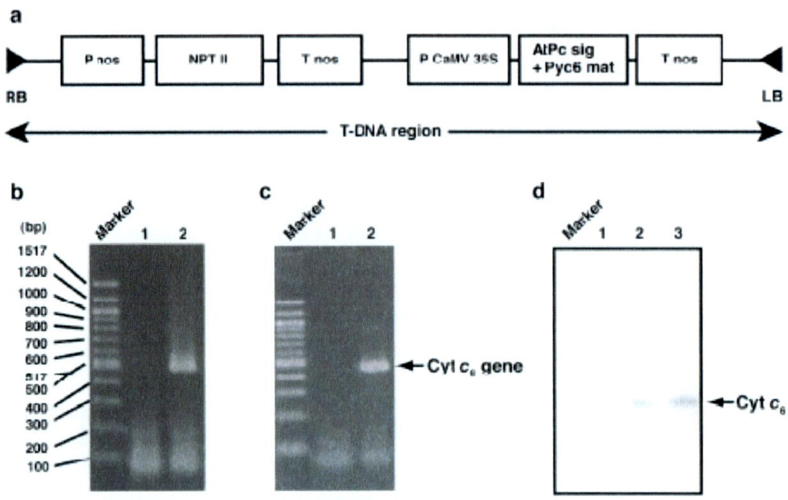

Figure 4. Construction of *Porphyra* cytochrome c_6 transformed *Arabidopsis*.

a) Construct of plasmid DNA. P nos, nopaline synthase promoter; NPT II, neomycin phosphotransferase; T nos, nopaline synthase terminator; P CaMV 35S, cauliflower mosaic virus 35S promoter; AtPc sig, *A. thaliana* plastocyanin transit peptides; Pyc6 mat, *P. yezoensis* cytochrome c_6 mature peptides; LB, left T-DNA border; RB, right T-DNA border.

b) Amplification of cytochrome c_6 gene in wild-type (1) and transgenic (2) plants using PCR.

c) RT-PCR analysis of wild-type (1) and transgenic (2) plants.

d) Western blot analysis of chloroplast proteins of wild-type (1) and transgenic (2) plants, and *P. yezoensis* (3).

Total RNA and poly $(A)^+$ mRNA were isolated from leaves of Cyt c_6-gene-positive T1 transformant and wild-type *Arabidopsis* plants. Two oligonucleotide primers, AtPcsig1 and Pyc62, were used in a reverse transcriptase PCR reaction to amplify and detect the mRNA transcript of the Cyt c_6 gene. The transgenic plants that harbored the Cyt c_6 gene exhibited a full-length Cyt c_6 mRNA transcript that corresponds to a theoretical length of 490 base pairs (Figure 4c). The expression of Cyt c_6 in the chloroplast of *Arabidopsis* could be detected by western blot analysis (Robinson, 1983) of chloroplast protein extracts. Cyt c_6 purified from *P. yezoensis* was used as a positive control. A 9.6 kDa protein that cross-reacted with Cyt c_6 antiserum was present in the T1 transgenic *A. thaliana* or *P. yezoensis* but was absent in the wild-type control (Figure 4d). In the chloroplast of transgenic *Arabidopsis*, the level of Cyt c_6 corresponds to 1.2-fold of that of PC when compared in molar ratio (Navarro et al., 2004; Kieselbach et al., 1998). The secondary transgenic plants (T2) that were harvested from the T1 transformants were used for further analysis. Approximately 90% of seeds harvested from the T1 transformants were Cyt c_6-gene-positive and the T2 transformants showed the same experimental results as the T1 transformants. The integration of T-DNA was detected by thermal asymmetric interlaced (TAIL)-PCR (Liu et al., 1995; Liu et al., 2005) revealed its location at approximately 13.7 Mbp from top of chromosome 3 in *A. thaliana* of mutant line A. There are no intrinsic genes.

To examine the change in the growth of transgenic plant expressing the *Porphyra* Cyt c_6 gene, we measured plant height, leaf and root lengths and the number of flowers using twenty wild-type and transgenic plants. The changes in growth were measured every 10 days, from 10 to 90 after cultivating. The height of the transgenic plants (T2) expressing Cyt c_6 was significantly larger than that of the wild-type plants from 30 to 80 days after planting (*t*-test, $P<0.05$) (Figure 5a). In particular, at 40 and 60 days after planting, the transgenic plants showed 1.9- and 1.3-fold increases in height, respectively, compared with the wild-type plants (Figs. 5a, 5c and 5e). The hypocotyl elongation of the transgenic plants also started seven to ten days earlier than that of the wild-type plants. As shown in Figures 5b, 5d and 5f, increased leaf length was also observed in the transgenic plants. The leaf length of the transgenic plants was significantly larger than that of the wild types from 40 to 70 days after planting (*t*-test, $P<0.05$). Final levels of both the plant height and leaf length were the same between the wild types and the transgenic plants. The root lengths of the transgenic *Arabidopsis* plants grown hydroponically were significantly larger than those of the wild-type plants from 30 to 70 days after planting (*t*-test, $P<0.05$) (Figure 6a). The 1.4- and 1.3-fold increase of the root lengths was observed on 40 and 60 days after planting, respectively (Figs. 6a, 6b and 6c). The final level of the root length on 80 and 90 days after planting was not different between the transgenic plants and the wild-type plants. Moreover, there were no significant differences in plant height, leaf and root lengths between the mutant line A and B. An average total numbers of flowers per one plant were 60 ± 2 for the wild-type *Arabidopsis*; 62 ± 3 for the transgenic plants, showing no difference between the wild-type and the transgenic plants. Flowering occurred 50 days after planting in the wild-type plants, whereas it occurred more than seven to twelve days earlier in the transgenic plants (data not shown). These results indicate that the expression of *Porphyra* Cyt c_6 in *A. thaliana* promotes the growth of plants in early stage.

To investigate whether the amount of metabolites in transgenic plants increased in comparison with that in wild-type plants, we measured the amounts of chlorophylls (Schreiber, 1986), proteins (Lowry et al., 1951), ATP (Larsson et al., 1979), NADPH (Häusler et al., 2000) and starch (Leegood, 1993). All experiments were carried out every ten days from 30 to 80 days after planting, using eight wild-type and transgenic plants (mutant line A) at 3hrs after illumination. Chlorophyll content in the transgenic plants expressed Cyt c_6 was 1.1-1.3-fold higher than that of wild-type plants (significant in *t*-test, $P<0.05$) (Figure 7a). Concomitantly, protein content of the transgenic plants was also significantly higher than that in the wild-type plants (*t*-test, $P<0.05$) (Figure 7b). The contents of ATP, NADPH and starch were also higher in the transgenic plants from 40 to 80 days after planting (Figs. 7c, 7d and 7e). For example, 60 days after planting, in the transformants, the amounts of protein, ATP, NADPH and starch were 1.1-, 1.9-, 1.4-, and 1.2-fold those in the wild type plants, respectively. The results clearly indicate that the larger plant body of the transgenic plants is not the result of succulent growth but is the consequence of substantial growth.

The *Arabidopsis* expressing *Porphyra* Cyt c_6 not only enhances growth measured in terms of plant height (Figure 4), and leaf (Figure 5) and root lengths (Figure 6), but also increases the amount of chlorophylls, proteins, ATP, NADPH and starch (Figure 7). The increasing rate of photosynthesis in transgenic plants thus could lead to an increase of ATP and NADPH, which are products of photosynthetic electron transfer and photophosphorylation and universal energy currencies of living cells.

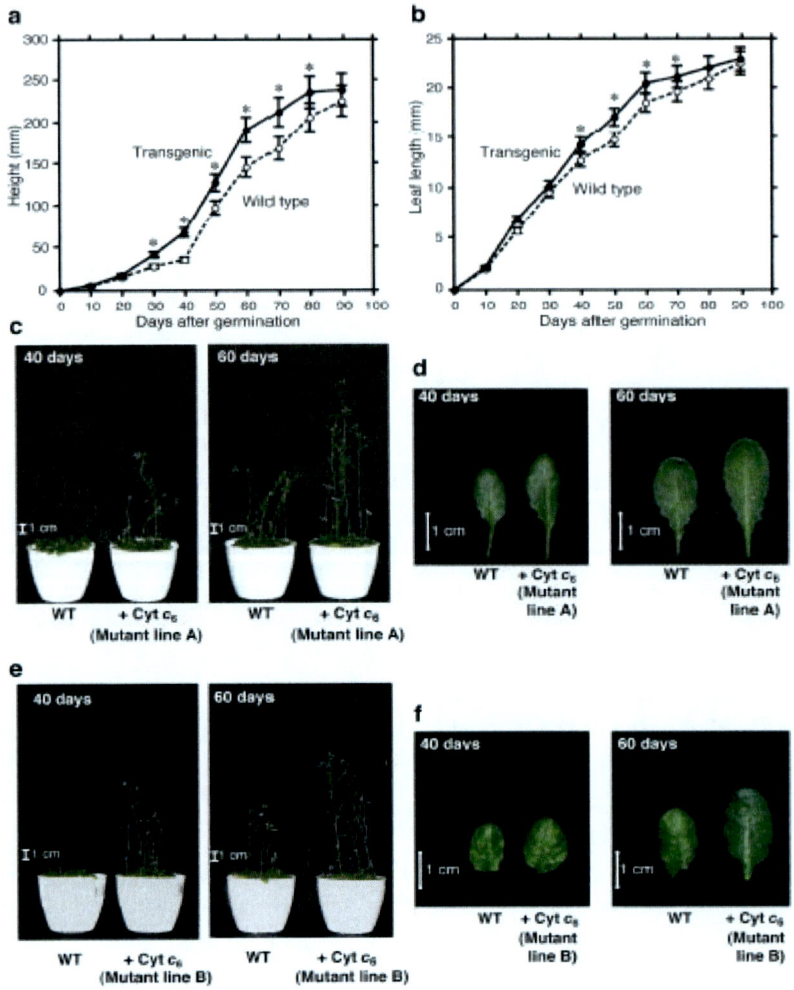

Figure 5 Plant growth.

a) Time courses of plant heights of wild-type (○) and transgenic (●) plants (mutant line A) in 360 ppm CO_2 at 50 μmol photons m^{-2} s^{-1}.

b) Time courses of leaf lengths of wild-type (○) and transgenic (●) plants (mutant line A).

c) Wild-type and transgenic *Arabidopsis* plants of mutant line A were photographed after 40 and 60 days.

d) The leaf lengths of wild-type and transgenic *Arabidopsis* plants of mutant line A were photographed after 40 and 60 days.

e) Wild-type and transgenic *Arabidopsis* plants of mutant line B were photographed after 40 and 60 days.

f) The leaf lengths of wild-type and transgenic *Arabidopsis* plants of mutant line B were photographed after 40 and 60 days.

The bars correspond to standard deviation from twenty plants in each. Asterisks indicate that the difference between wild-type and transgenic plants was significant by a t-test ($P<0.05$).

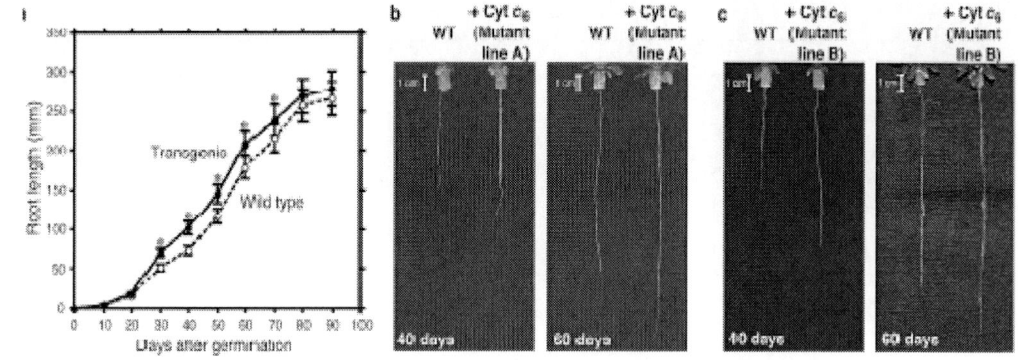

Figure 6 Development of root lengths.

Time courses of root lengths of wild-type (○) and transgenic (●) plants (mutant line A).

Root lengths of wild-type and transgenic *Arabidopsis* plants of mutant line A were photographed after 40 and 60 days.

Root lengths of wild-type and transgenic *Arabidopsis* plants of mutant line B were photographed after 40 and 60 days.

The bars and asterisks are indicated the same meaning as those in Figure 5.

In addition, since glutamine synthase, which plays pivotal roles in the biosynthesis of amino acids, catalyzes the ATP-dependent amination of glutamate to glutamine in plants (Temple et al. 1998), an increase in ATP levels of the transgenic plants may contribute to enhancement of the biosynthesis of amino acids and proteins. The content of starch and the CO_2 assimilation capacity of transgenic plants should be enhanced since increasing the contents of ATP and NADPH may be used for starch synthesis (Stark et al. 1992) and Calvin-Benson cycle. Furthermore, the higher plants Cyt c_{6A} cannot complement the PC-null *Arabidopsis* mutant phenotype (Weigel et al. 2003). Our results show that the algal Cyt c_6 transfers the electron directly to *Arabidopsis* PSI *in vitro* (Figure 3). Accordingly, introduction of the algal Cyt c_6 into the *A. thaliana* should be possible to complement the PC-null *Arabidopsis* mutant. This is the very interesting subject worthy of further study.

IV. ENHANCEMENT OF PHOTOSYNTHETIC ELECTRON TRANSFER IN TRANSGENIC *ARABIDOPSIS WITH* CYT *C6*

The capacity of CO_2 assimilation, which uses ATP and NADPH produced by photosynthetic electron transfer and photophosphorylation (Arnon 1984; Schurmann et al. 1971), was determined with the plants grown for 60 days after planting, which had fully developed leaves suitable for the measurements. The CO_2 assimilation capacity of the wild-type and the transgenic plants (mutant line A) was 1.420 ± 0.091 and 1.862 ± 0.304 (μmol CO_2 m^{-2} s^{-1}), respectively. The differences were also observed in plants grown for 70 or 80 days after planting, and statistically significant (*t*-test, $P<0.05$).

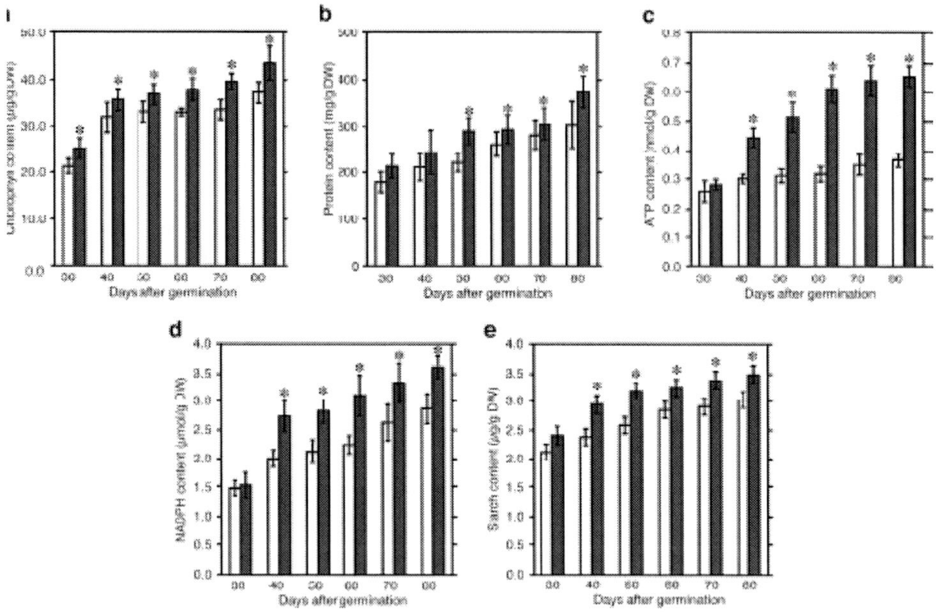

Figure 7 Comparison of the amount of metabolites between wild-type and transgenic plants grown for 30, 40, 50, 60, 70 and 80 days at 3hrs after illumination.

a) The content of chlorophylls.

b) The content of proteins.

c) The content of ATP.

d) The content of NADPH.

e) The content of starch.

Data are means ± standard deviations of eight plants in each experiment. The bars correspond to standard deviation from eight plants in each experiment. White and black boxes indicate wild-type and transgenic plants, respectively. Asterisks indicate that the difference between the wild-type and transgenic plants was significant by a t-test ($P<0.05$).

To investigate the cause of enhanced photosynthesis in transgenic plants expressing Cyt c_6, we determined photosynthetic parameters by measuring chlorophyll fluorescence under growth light condition using mutant line A plants. The parameter Φ_{PSII}, that reflects effective quantum yields of photosynthesis was somewhat higher in the transgenic plants (transgenic 0.716 ± 0.0077, wild-type 0.699 ± 0.0168), confirming that the expression of Cyt c_6 enhances photosynthesis. The increased quantum yield of photosynthesis must be due to either the increase of the quantum yield of open PSII (reflected in the parameter Fv'/Fm' (transgenic 0.793 ± 0.0121, wild-type 0.793 ± 0.0081)) or the decrease of excitation pressure 1-qP (transgenic 0.060 ± 0.0072, wild-type 0.042 ± 0.0020), which reflects the redox state of the plastoquinone pool of photosynthetic electron transport system. Fv'/Fm' (transgenic 0.744 ± 0.0125, wild-type 0.748 ± 0.0090), together with Fv/Fm, the maximum quantum yield of PSII, did not show any significant difference between wild-type and transgenic plants. On the other hand, 1-qP is lower in the transgenic plants by 30%, suggesting that the plastoquinone pool is more oxidized in the transgenic plants (*t*-test, $P<0.05$). These results indicate that an increase in the efficiency of photosynthetic electron transport is due to more efficient electron transfer in the downstream of plastoquinone pool but not to the change in PSII reaction center complexes. The parameter qN (transgenic 0.195 ± 0.0275, wild-type 0.194 ± 0.0348), which reflects non-photochemical quenching of chlorophyll fluorescence, did not show any difference between wild-type and transgenic plants.

Analysis of chlorophyll fluorescence indicate that the rate of linear electron transport in a steady state is enhanced in the transgenic plants compared with that in the wild-type plants and plastoquinone pool (1-qP) is more oxidized in transgenic *A. thaliana* plants than that in the wild-type plants. Generally, the rate-limiting step of the photosynthetic electron transport system is considered to be the electron transfer between reduced plastoquinone and Cyt b_6f (Junge, 1977). However, the expression of Cyt c_6, which is located downstream of Cyt b_6f in the photosynthetic electron transport, appears to lead to an increase in electron transfer efficiency. It was reported that increase in the relative amount of PSI leads to the increase in the rate of overall electron transfer in photosynthesis in the cyanobacterium *Synechococcus* sp. PCC6803 (Sonoike et al., 2001). This indicates that the enhancement of photosynthetic electron transfer could be realized by the increase in the rate of electron transfer even if it is not the rate-limiting step. The transgenic plants exhibited a higher Φ_{PSII} than that in the wild-type plants. The difference would be large enough to enhance the growth and increase the amounts of chlorophylls, proteins and metabolites such as ATP and NADPH, probably because the increase in electron transfer efficiency in the transgenic plants accumulates continuously during the entire growth period of the plants. Since the molar content of Cyt c_6 in the transgenic plant is estimated as 1.2-fold of that of PC, the total amount of electron donor to PSI in the transgenic plants should be more than twice of that in wild type plants, suggesting that this increase may primarily contribute to the increased rate of photosynthesis in the transgenic plants. Intrinsic PC is also expressed in the transgenic *Arabidopsis* plants. Intrinsic PC with higher affinity to PSI may compete with introduced Cyt c_6 with lower affinity. Therefore, it is possible that the expression of Cyt c_6 in PC-null *Arabidopsis* mutant would further enhance not only photosynthetic electron transfer but also growth and the amounts of the metabolites.

Here, we show that the enhancement of photosynthesis, metabolites and growth of the model plant *A. thaliana* could be realized by expressing the gene of a protein that functions in photosynthetic electron transport of the light reaction.

V. Search For Novel Algal Cyts *c6*

Here, the brown alga Hizikia fusiformis (Akazaki et al., 2008), and the diatom *Phaedactylum tricornutum* (Akazaki et al., 2009) show the ability to have physicochemical properties and 3D structure for biotechnological application of algal genes. Although chloroplasts were thought to have evolutionally arisen from cyanobacteria (Aitken, 1976), there are differences in the expression and genome coding of cytochrome c_6 between green and red algae. In the green alga *Chlamydomonas reinhardtii*, the gene for cytochrome c_6 exists in the genomic DNA and its coding region is interrupted by two introns (Hill et al.,

1991). On the other hand, in the red alga *Porphyra purpurea*, the petJ gene encoding cytochrome c_6 exists in the chloroplast genome (Reith and Munholland, 1993).

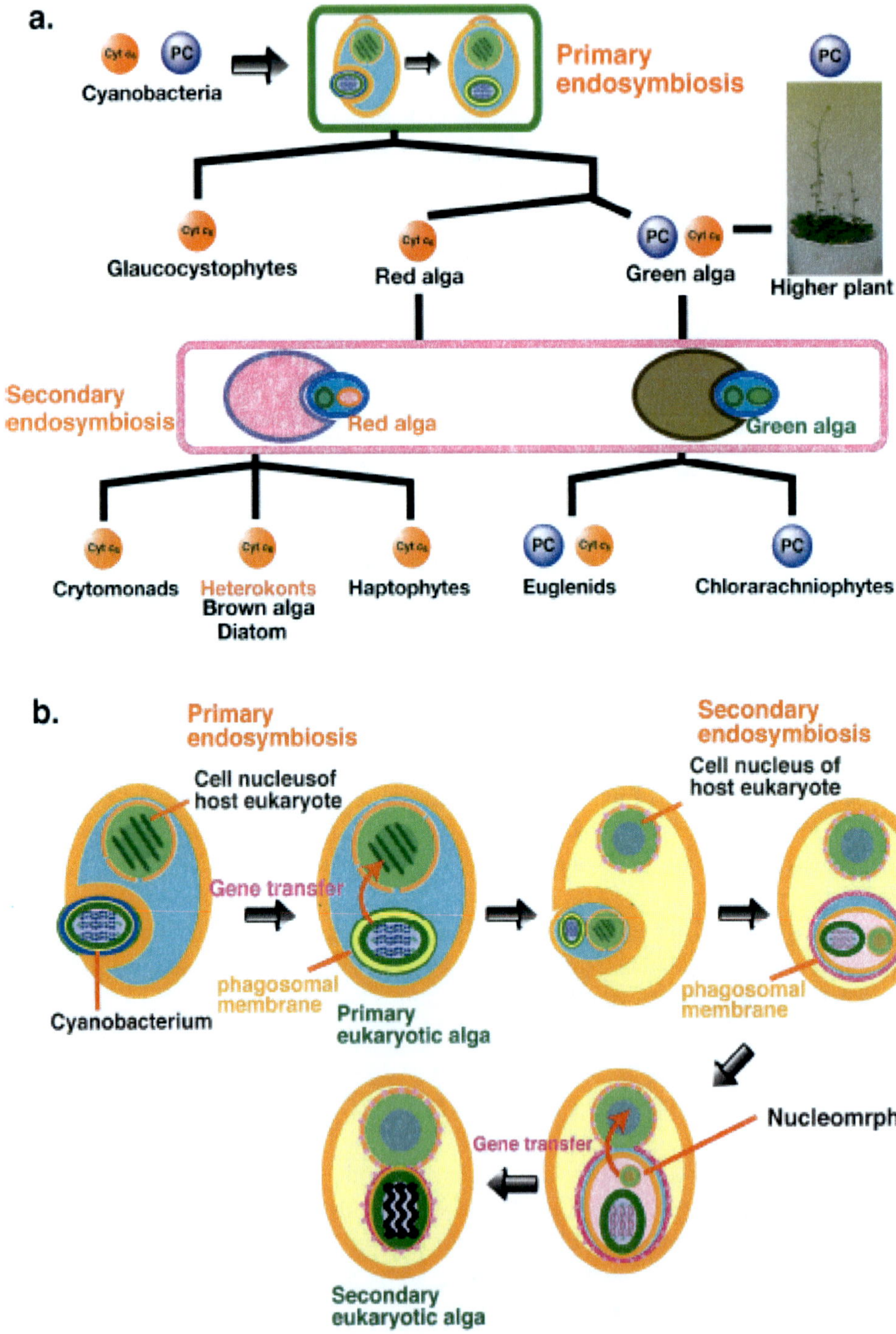

Figure 8. An estimated algal evolution (a) and endosymbiosis (b).

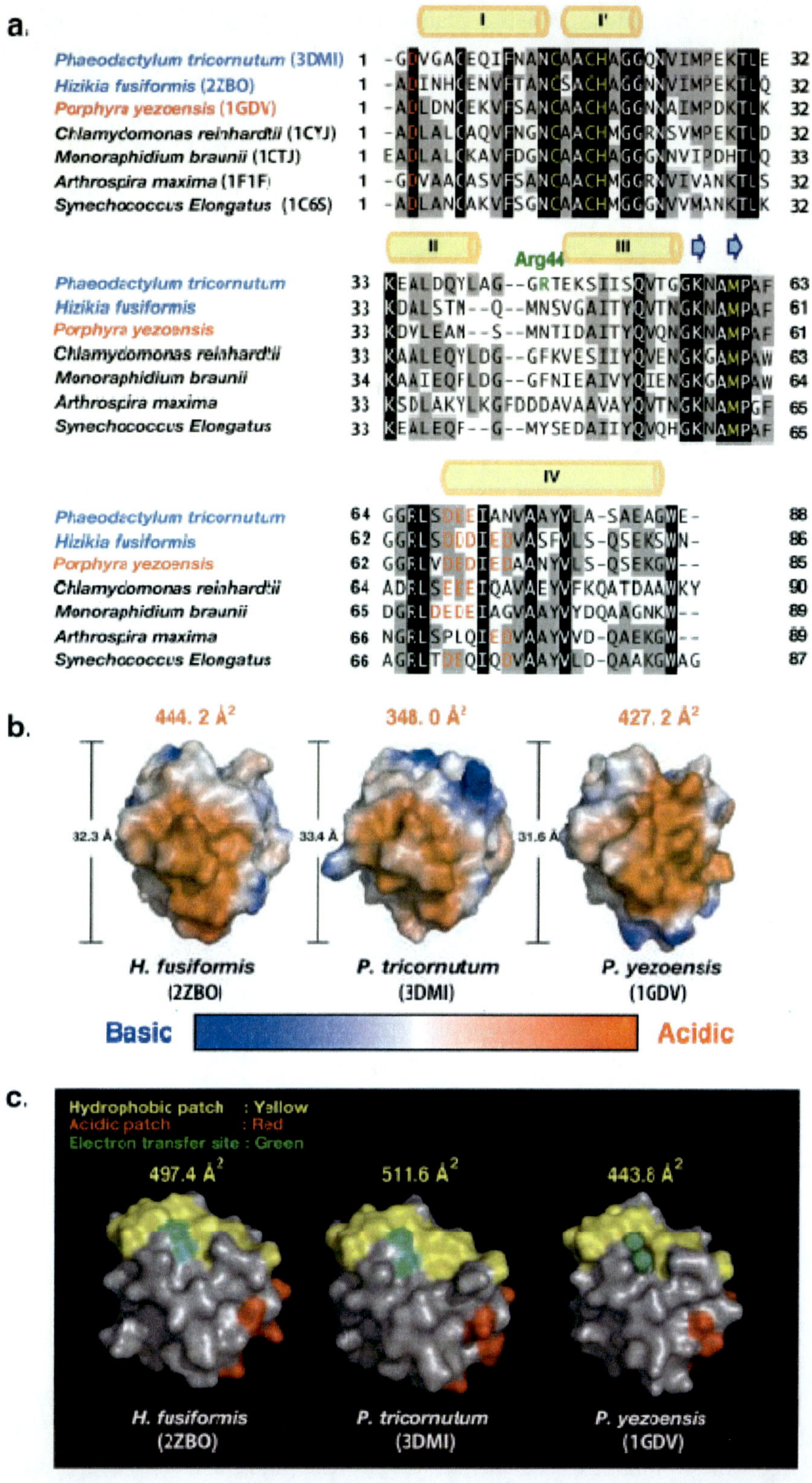

Figure 9. Amino acid sequence (a), and acidic (b) and hydrophobic (c) patch areas of Cyts c_6.

In eukaryotic brown algae, which contain no plastocyanin, the photosynthetic electron transport between cytochrome f and photosystem I is performed by only cytochrome c_6. It has generally been considered that brown algae acquired their chloroplasts via secondary endosymbiosis (Figure 8) involving a primitive red algal endosymbiont and a non-photosynthetic eukaryote host (Cavalier-Smith, 2000; McFadden, 1999). Although the physicochemical properties and amino acid sequence of cytochrome *c*6 from the brown algae Petalonia fascia and Alaria esculenta have been determined (Sugimura et al., 1981; Laycock, 1975), amino acid sequence of brown algal cytochrome c6 remains to be studied. Here, we determined amino acid sequence (Figure 9a) of algal cytochrome c6 from the brown alga *Hizikia fusiformis* (Akazaki et al., 2008) and from the diatom *Phaedactylum tricornutum* (Akazaki et al., 2009), which have been evolutionally eliminated from red algae, determined crystal structure of the proteins, and compared them with those of a red algal *Porphyla yezoensis* cytochromes *c*6 (Yamada et al., 2000).

Cytochrome *c*6 and plastocyanin have completely different amino acid sequences, and secondary and tertiary structures, but they constitute similar acidic and hydrophobic patches on their surface for the recognition of their interaction partners (Frazão et al., 1995; Ullmann et al., 1997). Area of the acidic and hydrophobic patches, and the redox potential of cytochrome c6 of (1) the brown alga *Hizikia fusiformis* (Akazaki et al., 2008), (2) the diatom *Phaedactylum tricornutum* (Akazaki et al., 2009), and (3) the red *Porphyla yezoensis* (Yamada et al., 2000) were (1) 444.2 Å, 497.4 Å and 313 mV, (2) 348.0 Å, 498.0 Å and 349 mV, and (3) 427.2 Å, 443.8 Å and 371 mV, respectively (Figures 9b and 9c). We suggest that both cytochromes c6 of *Hizikia fusifor*mis and *Phaedactylum tricornutum* may be worth evaluating further for an attempt to transform their genes into plant, because their properties are highly similar to that of the red alga *Porphyla yezoensis*.

CONCLUSION

Photosynthesis is the process by which higher plants, algae, and certain species of bacteria transform and store solar energy in the form of energy-rich organic molecules.

These compounds are, in turn, used as the energy source for all growth and reproduction in these and almost all other organisms. As such, virtually all life on the planet ultimately depends on photosynthetic energy conversion.

Design of hybrid and transgenic plants by genetic engineering, clean energy technology, food production and use of biomass as fuel, and alleviation of global warming by CO_2 sequestration with plants, algae and photosynthetic cyanobacteria are real current areas in photosynthesis research.

"Biotechnological application of *Porphyla yezoensis* genes" of this chapter is the first successful example to demonstrate enhancement of both the light and the dark reactions of photosynthesis and growth by introducing the gene of a red algal *Porphyla yezoensis* into the model plant *Arabidopsis thaliana*, based on the evolutionary history of algae to land plants and binding site of 3D structures for the photosynthetic electron transport protein, cytochrome c_6, and cytochrome *f* or photosystem I in chloroplast. Hence, development of biotechnological application of these findings to enhancement of the light and the dark reactions of photosynthesis and growth of plants would be needed to further elucidate 3D structures and information of detailed thylakoid membrane proteins, which were considered evolutionary history, and to attempt to introduce their genes into various kinds of land plants.

Acknowledgments

We thank Aiko Nakazawa, Ikuko Shirasaki, Harumi Nakade, Mao Hirayama, Shouhei Kojima, Kouyou Sekine and Wataru Hiruma for their technical assistance. This work was supported in part by a Nihon University Multidisciplinary Research Grant for 2009.

References

Aitken, A. (1976). Protein evolution in cyanobacteria. *Nature*, *263*, 793-796.

Akazaki, H. Kawai, F. Chida, H. Matsumoto, Y. Hirayama, M. Hoshikawa, K. Unzai, S. Hakamata, W. Nishio, T. Park, S. Y. and Oku, T. (2008). Cloning, expression and purification of cytochrome c_6 from the brown alga *Hizikia fusiformis* and complete X-ray diffraction analysis of the structure. *Acta Cryst.*, *F64*, 674-680.

Akazaki, H. Kawai, F. Hosokawa, M. Hama, T. Chida, H. Hirano, T. Lim, B. K. Sakurai, N. Hakamata, W. Park, S. Y. Nishio, T. and Oku, T. (2009). Crystallization and structural analysis of cytochrome c_6 from the diatom *Phaeodactylum tricornutum* at 1.5 Å resolution. *Biosci. Biotechnol. Biochem.*, *73*, 189-191.

Arnon, D. I. (1984). The discovery of photosynthetic phosphorylation. *Trends Biochem. Sci.*, *9*, 258-262.

Cavalier-Smith, T. (2000). Membrane heredity and early chloroplast evolution. *Trends Plant Sci.*, *5*, 174-182.

Chida, H. Nakazawa, A. Akazaki, H. Hirano, T. Suruga, K. Ogawa, M. Satoh, T. Kadokura, K. Yamada, S. Hakamata, W. Isobe, K. Ishii, R. Ito, T. Nishio, T. Sonoike, K. and Oku, T. (2007). Expression of the algal cytochrome c_6 gene in *Arabidopsis* enhances photosynthesis and growth. *Plant Cell Physiol.*, *48*, 948-957.

Patent: WO 2006/082992 A1 "Method of constructing novel higher plant and method of promoting the growth of higher plant" (Nihon University, Japan).

Chida, H. Yokoyama, T. Kawai, F. Nakazawa, A. Akazaki, H. Takayama, Y. Hirano, T. Suruga, K. Satoh, T. Yamada, S. Kawachi, R. Unzai, S. Nishio, T. Park, SY. and Oku, T. (2006). Crystal structure of oxidized cytochrome c_{6A} from *Arabidopsis thaliana*. *FEBS Lett.*, *580*, 3763-3768.

Dandekar, A ; Fisk, H J. Plant transformation: Agrobacterium-mediated gene transfer. In: Peña L, editor. *Methods in Molecular Biology*: *Transgenic Plants Methods and Protocols*. Totowa, New Jersey: Humana Press, Inc., 2005; vol. 286, 35-46.

De la Rosa, M. A. Navarro, J. A. Díaz-Quintana, A. De la Cerda, B. Molina-Heredia, F. P. Balme, A. Murdoch-Pdel, S. Díaz-Moreno, I Durán, R. V. and Hervás, M. (2002). An evolutionary analysis of the reaction mechanisms of photosystem I reduction by cytochrome c_6 and plastocyanin. *Bioelectrochemistry, 55,* 41-45.

De la Rosa, M. A. Molina-Heredia, F. P. Hervás, M. and Navarro, J. A. (2006). *Photosystem I: The light-driven plastocyanin.* Netherlands: Springer.

Finazzi, G. Sommer, F. and Hippler, M. (2005). Release of oxidized plastocyanin from photosystem I limits electron transfer between photosystem I and cytochrome b_6f complex in *vivo*. *Proc. Natl. Acad. Sci. USA, 102*, 7031-7036.

Frazão, C. Soares, C. M. Carrondo, M. A. Pohl, E. Dauter, Z. Wilson, K. S. Hervás, M. Navarro, J. A. De la Rosa, M. A. and Sheldrick, G. M. (1995). Ab initio determination of the crystal structure of cytochrome c_6 and comparison with plastocyanin. *Structure*, *3*, 1159-1169.

Gupta, R. He, Z. and Luan, S. (2002). Functional relationship of cytochrome c_6 and plastocyanin in *Arabidopsis*. *Nature, 417*, 567-571.

Hageman, J. Robinson, C. Smeekens, S. and Weisbeek, P. (1986). A thylakoid processing protease is required for complete maturation of the lumen protein plastocyanin. *Nature*, *324*, 567-569.

Häusler, R. E. Fischer, K. L. and Flügge, U. F. (2000). Determination of low-abundant metabolites in plant extracts by NAD(P)H fluorescence with a microtiter plate reader. *Anal. Biochem., 281*, 1-8.

Hervás, M. Navarro, J. A. Molina-Heredia, F. P. and De la Rosa, M. A. (1998). The reaction mechanism of photosystem I reduction by plastocyanin and cytochrome c_6 follows two different kinetic models in the cyanobacterium *Pseudanabaena* sp. PCC 6903. *Photosynth. Res., 57*, 93-100.

Hervás, M. Ortega, J. M. Navarro, J. A. De la Rosa, M. A. and Bottin, H. (1994). Laser flash kinetic analysis of *Synechocystis* PCC 6803 cytochrome c_6 and plastocyanin oxidation by photosystem I. *Biochim. Biophys. Acta, 1184*, 235-241.

Hill, K. L. Li, H. H. Singer, J. S. and Merchant, S. (1991). Isolation and structural characterization of the *Chlamydomonas reinhardtii* gene for cytochrome c_6. Analysis of the kinetics and metal specificity of its copper-responsive expression. *J. Biol. Chem*., *226*, 15060-15067.

Hiyama, T. (2004). *Methods in Molecular Biology, vol. 274*: *Photosynthesis Research Protocols*. Totowa, New Jersey: Humana Press, Inc.

Howe, C. J. Schlarb-Ridley, B. G. Wastl, J. Purton, S. and Bendall, D. S. (2006). The novel cytochrome c_6 of chloroplasts: a case of evolutionary *bricolage*? *J. Exp. Bot*., *57*, 13-22.

Junge, W. (1977). *Encyclopedia of Plant physiology* Vol. 5: *Photosynthesis.* New York: Springer-Verlag.

Katoh, S. (1960a). Crystallization of an algal cytochrome, *Porphyra tenera*-cytochrome 553. *Nature*, *186*, 138-139.

Katoh, S. (1960b). A New Copper Protein from *Chlorella Ellipsoidea. Nature*, *186*, 533-534.

Katoh, S. and Takamiya, A. (1961). A new leaf copper protein "plastocyanin," a natural Hill oxidant. *Nature*, *189*, 665-666.

Kerfeld, C. A. and Krogman, D. W. (1998). Photosynthetic cytochromes *c* in cyanobacteria, algae, and plants. *Annu. Rev. Plant Physiol. Plant Mol. Biol.*, *49*, 397- 425.

Kieselbach, T. Hagman, A. Anderson, B. and Schröder, W. P. (1998). The thylakoid lumen of chloroplasts. *J. Biol. Chem.*, *273*, 6710-6716.

Ku, M. S. Agarie, S. Nomura, M. Fukayama, H. Tsuchida, H. Ono, K. Hirose, S. Toki, S. Miyao, M. and Matsuoka, M. (1999). High-level expression of maize phosphoenolpyruvate carboxylase in transgenic rice plants. *Nature Biotechnol.*, *17*, 76-80.

Larsson, C. M. and Olsson, T. (1979). Firefly assay of adenine nucleotides from algae: Comparison of extraction methods. *Plant Cell Physiol.*, *20*, 145-155.

Laycock, M. V. (1975). The amino acid sequence of cytochrome *f* from the brown alga *Alaria esculenta* (L.) Grev. *Biochem. J.*, *149*, 271-279.

Leegood, R. C. (1993). *Photosynthesis and production in a changing environment: A field and laboratory manual.* London: Chapman and Hall.

Liu, Y.G. Mitsukawa, N. Oosumi, T. and Whittier, R.F. (1995). Efficient isolation and mapping of Arabidopsis thaliana T-DNA insert junctions by thermal asymmetric interlaced PCR. *Plant J.*, *8*, 457–463.

Liu, Y.G. Chen, Y. and Zhang, Q. (2005). Amplification of genomic sequences flanking T-DNA insertions by thermal asymmetric interlaced polymerase chain reaction. *Methods Mol. Biol.*, *286*, 341–348.

Lowry, O. H. Rosebrough, N. J. Farr, A. L. and Randall, R. J. (1951). Protein measurement with the folin phenol reagent. *J. Biol. Chem.*, *193*, 265-275.

McFadden, G. I. (1999). Plastids and protein targeting. *J. Eukaryot. Microbiol.*, *46*, 339-346.

Merchant, S. and Bogorad, L. (1987). The Cu(II)-repressible plastidic cytochrome c: cloning and sequence of a complementary DNA for the pre-apoprotein. *EMBO J.*, *6*, 2531-2535.

Miyagawa, Y. Tamoi, M. and Shigeoka, S. (2001). Over-expression of a cyanobacterial fructose-1,6-/sedoheptulose-1,7-bisphosphatase in tobacco enhances photosynthesis and growth. *Nature Biotechnol.*, *19*, 965-969.

Molina-Heredia, F. P. Wastl, J. Navarro, J. A. Bendall, D. S. Hervás, M. Howe, C. J. and De La Rosa, M. A. (2003). A new function for an old cytochrome? *Nature*, *42*, 33-34.

Mackinney, G. (1941). Absorption of light by chlorophyll solutions. *J. Biol. Chem.*, *140*, 315-322.

Navarro, J A; Hervás, M; De la Rosa, M A. Purification of plastocyanin and cytochrome c_6 from plants, green algae, and cyanobacteria. In: Carpentier R, editor. *Methods in Molecular Biology: Photosynthesis Research Protocols*. Totowa, New Jersey: Humana Press, Inc., 2004; *vol. 274*, 79-92.

Ramiro, E. R. Anabella, L. Hugo, O. P. Matial, Z. Martine, P. Javier, F. P. Tognetti, V. B. Tschiersch, H. Hajirezaei, M. R. Valle, E. M. and Carrillo, N. (2007). Transgenic tobacco plants over-expressing chloroplastic ferredoxin-NADP(H) reductase display normal rates of photosynthesis and increased tolerance to oxidative stress. *Plant Physiol.*, *143*, 639-649.

Reith, M. and Munholland, J. (1993) A High-Resolution Gene Map of the Chloroplast Genome of the Red Alga *Porphyra purpurea*. *Plant Cell*, *5*, 465-475.

Robinson, S. P. (1983). Isolation of intact chloroplasts with high CO_2 fixation capacity from sugar beet leaves containing calcium oxalate. *Photosynth. Res.*, *4*, 281-287.

Satoh, T. Itoga, A. Isogai, Y. Kurihara, M. Yamada, S. Natori, M. Suzuki, N. Suruga, K. Kawachi, R. Arahira, M. Nishio, T. Fukazawa, C. and Oku, T. (2002). Increasing the conformational stability by replacement of heme axial ligand in *c*-type cytochrome. *FEBS Lett.*, *531*, 543-547.

Schurmann, P. Buchanan, B. B. and Arnon, D. I. (1971). Role of cyclic photophosphorylation in photosynthetic carbon dioxide assimilation by isolated chloroplasts. *Biochim. Biophys. Acta*, *267*, 111-124.

Schreiber, U. (1986). Detection of rapid induction kinetics with a new type high-frequency modulated chlorophyll fluorometer. *Photosynth. Res.*, *9*, 261-272.

Sigfridsson, K. (1998). Plastocyanin, an electron-transfer protein. *Photosynth. Res.*, *57*, 1-28.

Soll, J. and Schleiff, E. (2004). Protein import into chloroplasts. *Nat. Rev. Cell. Biol.*, *5*, 198-208.

Sonoike, K. Hihara, Y. and Ikeuchi, M. (2001). Physiological Significance of the regulation of photosystem stoichiometry upon high light acclimation of *Synechocystis* sp. PCC6803. *Plant Cell Physiol.*, *42*, 379-384.

Stark, D. M. Timmerman, K. P. Barry, G. F. Preiss, J. and Kishore, G. M. (1992). Regulation of the amount of starch in plant tissues by ADP-glucose pyrophosphorylase. *Science*, *9*, 287-292.

Sugimura, Y. Hase, T. Matsubara, H. and Shimokoriyama, M. (1981). Studies on algal cytochromes. III. Amino acid sequence of cytochrome *c*-553 from a brown alga, *Petalonia fascia*. *J. Biochem.*, *90*, 1213-1219.

Temple, S. J. Vance, C. P. and Gantt, J. S. (1998). Glutamate synthase and nitrogen assimilation. *Trends Plant Sci.*, *3*, 51-56.

Theologis, A. Ecker, J. R. Palm, C. J. Federspiel, N. A. Kaul, S. White, O. Alonso, J. Altafi, H. Araujo, R. Bowman, C. L. Brooks, S. Y. Buehler, E. Chan, A. Chao, Q. Chen, H. Cheuk, R. F. Chin, C. W. Chung, M. K. Conn, L. Conway, A. B. Conway, A. R. Creasy, T. H. Dewar, K. Dunn, P. Etgu, P. Feldblyum, T. V. Feng, J. Fong, B. Fujii, C. Y Gill, J. E. Goldsmith, A. D. Haas, B. Hansen, N. F. Hughes, B. Huizar, L. Hunter, J. L. Jenkins, J. Johnson-Hopson, C. Khan, S. Khaykin, E. Kim, C. J. Koo, H. L. Kremenetskaia, I. Kurtz, D. B. Kwan, A. Lam, B. Langin-Hooper, S. Lee, A. Lee, J. M. Lenz, C. A. Li, J. H. Li, Y. Lin, X. Liu, S. X. Liu, Z. A. Luros, J. S. Maiti, R. Marziali, A. Militscher, J. Miranda, M. Nguyen, M. Nierman, W. C. Osborne, B. I. Pai, G. Peterson, J. Pham, P. K. Rizzo, M. Rooney, T. Rowley, D. Sakano, H. Salzbergm, S. L. Schwartz, J. R. Shinn, P. Southwick, A. M. Sun, H. Tallon, L. J. Tambunga, G. Toriumi, M. J. Town, C. D. Utterback, T. Van, Aken. S. Vaysberg, M. Vysotskaia, V. S. Walker, M. Wu, D. Yu, G. Fraser, C. M. Venter, J. C. and Davis, R. W. (2000). Sequence and analysis of chromosome 1 of the plant *Arabidopsis thaliana*. *Nature*, *408*, 816-820.

Ullmann, G. M. Hauswald, M. Jensen, A. Kostic, N. M. and Knapp, E. W. (1997). Comparison of the physiologically equivalent proteins cytochrome c_6 and plastocyanin on the basis of their electrostatic potentials. Tryptophan 63 in cytochrome c_6 may be isofunctional with tyrosine 83 in plastocyanin. *Biochemistry*, *36*, 16187-16196.

Wastl, J. Bendal, D. S. and Howe, C. (2002). Higher plants contain a modified cytochrome c_6. *Trends Plant Sci.*, *7*, 244-245.

Weigel, M. Varotto, C. Pesaresi, P. Finazzi, G. Rappaport, F. Salamini, F. and Leister, D. (2003). Plastocyanin is indispensable for photosynthetic electron flow in *Arabidopsis thaliana*. *J Biol. Chem.*, *278*, 31286-31289.

Yamada, S. Park, S. Y. Shimizu, H. Koshizuka, Y. Kadokura, K. Satoh, T. Suruga, K. Ogawa, M. Isogai, Y. Nishio, T. Shiro, Y. and Oku, T. (2000). Structure of cytochrome c_6 from the red alga *Porphyra yezoensis* at 1.57 Å resolution. *Acta Cryst. D*, *56*, 1577-1582

PERMISSION TO REUSE JOURNAL

We obtained permission to reuse *Plant Cell Physiology* (2007), Vol. 48, 948-957p; "Expression of the Algal Cytochrome *c6* Gene in *Arabidopsis* Enhances Photosynthesis and Growth" by permission of Oxford University Press, and to reuse *Acta Crystallographica* (2008), Section F, 674-680p; "Cloning and crystal structure of cytochrome c_6 from the brown alga *Hizikia fusiformis*" by permission of International Union of Crystallography.

In: Porphyra yezoensis
Editor: Koji Mikami

ISBN 978-1-61122-304-0

Chapter 10

NUTRITIONAL AND PHYSIOLOGICAL FUNCTIONS OF *PORPHYRA*

Tatsuya Sugawara, Shota Sakai and Takashi Hirata[1]
Graduate School of Agriculture, Kyoto University,
Kyoto 606-8502, Japan

ABSTRACT

The genus *Porphyra,* including *P. tenera* and *P. yezoensis*, is an important type of sea algae, which is widely consumed in East Asia including Japan, China and Korea. Dried nori prepared from *Porphyra* spp. contains many nutritional and functional components such as proteins, dietary fibers, carotenoids, polyunsaturated fatty acids (PUFA) and mycosporine-like amino acids (MAAs). Many studies have shown the beneficial effects of nori and their components, including anti-inflammatory, antioxidative and anti-tumor activities, and the improvement of lipid metabolism. Daily consumption of *Porphyra* should contribute to human health via its various nutritional and physiological functions.

INTRODUCTION

Marine red algae of the genus *Porphyra* are abundantly cultivated in East Asia. Nori prepared from *Porphyra* spp., generally *P. yezoensis,* is widely consumed as a familiar daily food in Japan. Dried nori sheets are obtained through a process that involves washing the raw laver roughly with water, cutting it coarsely, and drying it. It has long been believed that nori is a healthy food and is very stable to oxidative stress during storage under dry conditions. *Porphyra* spp. contains many nutritional components such as proteins, dietary fibers, minerals, carotenoids, and polyunsaturated fatty acids (PUFAs). Thus, dietary nori contributes to the importance of essential elements in the diet. In particular, phycoerythrin, porphyran and porphyra-334 are present in *Porphyra* spp. but are not found in terrestrial animals and plants.

1 Corresponding author: Takashi Hirata, Phone : +81-(0)75-753-6211, E-mail : hiratan@kais.kyoto-u.ac.jp.

In this chapter, we review the nutritional and physiological functions of components that are distinctive of *Porphyra* spp., including *P. yezoensis*.

NUTRITIONAL AND FOOD FUNCTIONS OF DIETARY NORI

Nori is a traditional Japanese marine foodstuff generally prepared from *P. yezoensis*. Dried nori is well known to contain a large amount of proteins (~44%), dietary fibers (~44%) and minerals (~10%) (Morita and Tobiishi, 2002). Edible marine algae are well known as food supplements to help meet the recommended daily intake of some essential minerals (Rupérez, 2002). Minerals found in *Porphyra* spp. include macro elements (Na, K, Ca, Mg, and P), trace elements (Fe, Mn, Zn, Cu, Se, and I) and ultra-trace elements (As, Pb, Cd, and Hg) (Dawczynski et al., 2007). Algae, including *Porphyra* spp., are also suitable sources of bio-available iron and could help to improve iron nutrition (García-Casal, 2007). Generally, the contents of ultra-trace elements (heavy metals) in nori, such as Pb, Cd, and Hg, which have a toxicological potential, can be assessed as harmless (Dawczynski et al., 2007).

A substantial amount of vitamin B_{12} is found in dried nori and is at a markedly higher level than in other algae (Watanabe et al., 1999; Miyamoto et al., 2009). Early studies reported that algal vitamin B_{12} appears to be inactive for mammals, because most vitamin B_{12} in algae exists as vitamin B_{12} analogues (Dagnelie et al., 1991; van den Berg et al., 1988). However, Rauma et al. (1995) reported that some algae can supply adequate amounts of bio-available vitamin B_{12} for strict vegetarians. Takenaka et al. (2001) showed that feeding dried-nori to vitamin B_{12}-deficient rats significantly improved their vitamin B_{12} status. Those results strongly suggest that the biologically active vitamin B_{12} compounds found in *Porphyra* spp. are also active in humans.

The nutritional and physiological effects of algae have received much attention during the past decades, and the anticancer effects of dietary algae have been reported *in vitro* (Yuan and Walsh, 2006; Okai et al., 1994) and *in vivo* (Yamamoto and Maruyama, 1985; Yamamoto et al., 1987). Noda et al. evaluated the antitumor activity of marine algae (Noda et al., 1990), and reported that oral administration of powdered *P. yezoensis* provided significant antitumor activity in mice implanted with Ehrlich carcinoma cells (53% inhibition). Appreciable levels of inhibition of Ehrlich carcinoma growth was due to porphyran, a dietary fiber prepared from *P. yezoensis*. A glycolipid and phospholipid fractions from *P. yezoensis* were also effective against the growth of Meth-A fibrosarcoma cells. In a case-control study among Korean women as an epidemiologic study, the daily intake of gim (*Porphyra* spp.) was inversely associated with the risk of breast cancer, but wakame (*Undaria pinnatifida*) was not (Yang et al., 2010). Their results suggested that consumption of *Porphyra* may reduce the risk of breast cancer and that the protective effect may be due to the antioxidant and antitumor activities of its components, such as dietary fibers, proteins, polyphenols, carotenoids, chlorophyll, etc.

Dioxin, which is one of the most toxic substances among synthetic chemicals, is a potent pollutant of the environment. Dioxin can be detected worldwide and is generated by incinerators when rubbish is destroyed by fire. Morita et al. reported that dietary *P. yezoensis* increased the fecal excretion of dioxin in rats (Morita and Tobiishi, 2002). They suggested that this effect might be due to the dietary fiber and chlorophyll in *P. yezoensis*.

Dietary nori (*P. tenera*) has been evaluated as a source of dietary fiber by some studies using rats (Goñi et al., 2001; Gudiel-Urbano and Goñi, 2002a; Gudiel-Urbano and Goñi, 2002 b). The intake of nori did not affect body weight gain or food intake. However, nori in the diet was associated with a significant increase of cecal, fresh stool and dry stool weight. Nori-fed animals showed a significantly lower apparent digestibility of protein and fat, but absorbed nitrogen was more effectively used (Gudiel-Urbano and Goñi, 2002a). Consumption of nori also caused a marked decrease in the fermentative capacity of intestinal microorganisms (Goñi et al., 2001) Cecal bacterial activities (β-glucuronidase, β-glycosidase, azoreductase, nitroreductase and nitrate reductase) were lower in the nori-fed animal group than in the cellulose-fed control group (Gudiel-Urbano and Goñi, 2002b). Adaptation to diets containing nori would be associated with changes in microbial activity that involve decreases in reductive and hydrolytic enzyme activities implicated in the conversion of procarcinogens into carcinogens.

PORPHYRAN

Porphyran, the sulfated polysaccharide comprising the hot-water soluble portion of the algal cell wall, is one of the main components of *Porphyra* spp. (Morrice et al., 1983). Porphyran is related to agarose in that it contains disaccharide units consisting of 3-linked β-D-galactosyl residues alternating with 4-linked 3,6-anhydro-α-L-galactose, but differs in that some residues occur as the 6-sulfate and the 6-O-methyl (Figure 1). It is the major dietary fiber in nori and usually accounts for nearly 40% of the total mass. Therefore, the nutritional and physiological functions of porphyran are important, and many studies have demonstrated its beneficial effects including its antioxidative, anti-tumor, immuno-modulating, and lipid metabolism improving activities.

It has been reported that sulfated polysaccharides from marine algae have antioxidant activities and porphyran from *P. haitanensis* was also shown to have antioxidant activity, including scavenging activity against superoxide and hydroxyl radicals (Zhang et al., 2003a). Zhang et al. (2009) reported that the low molecular weight fraction of porphyran prepared from *P. haitanensis* (~6,000 Da) has scavenging activity toward superoxide, hydroxyl and DPPH radicals. Interestingly, porphyrans degraded by free radicals show stronger antioxidant activities, and those activities increase with reduced molecular weight (Zhao et al., 2006). Carboxyl groups in the porphyran molecule may have an important effect on its antioxidant activities.

In addition, it has been reported that the intraperitoneal administration of porphyran significantly decreased lipid peroxidation in aging mice (Zhang et al., 2003b; Zhang et al., 2004). The total antioxidant capacity, as well as the activities of superoxide dismutase and glutathione peroxidase, increased in porphyran-treated mice. The effects of porphyran on the prevention of chemical acute liver injury have also been shown (Guo et al., 2007). In CCl_4-treated mice, an intraperitoneal injection of porphyran significantly reduced levels of alanine aminotransferase (ALT) and aspartate aminotransferase (AST) and the content of malondialdehyde in the liver. Porphyran may play a protective role against the hepatotoxicity induced by CCl_4 in mice, and its mechanism may be related to the free radical scavenging activities by increasing superoxide dismutase activities.

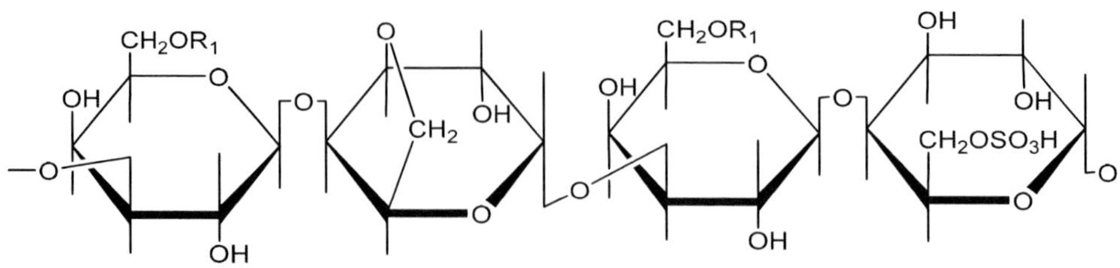

Figure 1. Structure of porphyran. R_1 = H or CH_3

The immunomodulating effect of porphyran has also been reported. Yoshizawa et al. demonstrated that two porphyran-rich fractions prepared from *P. yezoensis* directly activated murine macrophages in a cultured cell and intraperitoneal injection study (Yoshizawa et al., 1993). Later, they reported that digests of porphyran with β-agarase showed higher activity and lower viscosity than undigested porphyran (Yoshizawa et al., 1995). They evaluated the macrophage-stimulating activities of desulfated, sulfated, fractionated, and enzyme-degraded porphyran. Their results suggested that the sulfate groups in porphyran play a role in macrophage stimulation through direct interaction between the sulfate groups and macrophages or indirectly by maintaining the conformation of the active moieties. On the other hand, Ishihara et al. studied the effect of porphyran on the contact hypersensitivity reaction in mice to evaluate the anti-allergic activity (Ishihara et al., 2005). They found that oral administration of porphyran (2% in drinking water) suppressed the ear edema induced by 2,4,6-trinitrochlorobenzene due to suppression of serum IgE and IFN-γ production in the ear. Porphyran is a non-digestible dietary fiber, so it is unlikely that porphyran or its digestion products can interact directly with internal immune cells. It is possible that intestinal changes, such as elevating short chain fatty acid concentrations and lowering the pH in the cecum, affect the immune reaction. In a study of human gastric cancer cells, porphyran induced cell death in a dose-dependent manner, but did not affect the growth of normal cells (Kwon and Nam, 2006). The addition of 0.1% porphyran reduced DNA synthesis and insulin-like growth factor-I receptor (IGF-IR) phosphorylation, and increased caspase-3 activation, which induced apoptosis.

Recent studies have shown that dietary fiber is important for human health. Algae are a natural source of dietary fiber that differs from terrestrial plant fiber chemically and physico-chemically. Porphyran is the major dietary fiber in *Porphyra* spp.. A dietary supplement of nori (*P. tenera*) but not kombu (*Laminaria digitata*) decreased cholesterolemia in hypercholesterolemic rats (Bocanegra et al., 2008). The porphyran explains this effect at least in part because the ionic groups of molecules composing the fibers are able to sequester bile acids and consequently increase fecal bile acid. In rats fed a cholesterol-free diet, renal adipose tissue weights and serum cholesterol levels were significantly lower in the porphyran-fed group than in the cellulose-fed and the agar-fed groups (Tsuge et al., 2004). Fecal excretion of fatty acids and neutral sterols was markedly enhanced by porphyran ingestion, which suggests that porphyran has a potent effect to interfere with the intestinal absorption of cholesterol and fatty acids. In addition, porphyran reduces the secretion of apolipoprotein B100 (ApoB100) and lipid synthesis in human hepatoma HepG2 cells (Inoue et al., 2009). The lipid-lowering effect of dietary porphyran might be due not only to the suppression of intestinal absorption of lipids but also to direct effects on liver cells. It is important for understanding the effect of dietary porphyran to clarify the digestion and intestinal absorption, because dietary fibers are usually not absorbed from the digestive tract. We found that dietary porphyran ameliorates carbohydrate metabolism in KK-Ay mice, which are model animals for diabetes with obesity, by increasing levels of adiponectin (unpublished data). Daily consumption of porphyran might contribute to improve that metabolic syndrome, and the detailed mechanism of the effect should be further clarified.

Marine algae contain sulphated polysaccharides that are absent in terrestrial plants. Nori is the most important nutritional alga in Japan, and porphyran is the major polysaccharide in nori as noted above. Microbes in the gut supply the human body with energy from dietary polysaccharides through carbohydrate active enzymes, which are absent in the human genome. Most interestingly, comparative gut metagenome analyses show that porphyranase, which can degrade porphyran, is frequently found in the Japanese population but is absent in North Americans (Hehemann et al., 2010). Hehemann et al. (2010) concluded that porphyranase from a member of the marine Bacteroidetes, *Zobellia galactanivorans*, is active on porphyran and the gene encoding that it has been transferred to the gut bacterium *Bacteroides plebeius* isolated from Japanese individuals. Algae with associated marine bacteria may have been the route by which the carbohydrate active enzymes were acquired in human gut bacteria.

PORPHYRA-334

Mycosporine-like amino acids (MAAs), which are distributed widely throughout freshwater and marine organisms, have become of concern in terms of their UV-absorbing capacity (Shick and Dunlap, 2002). The family of MAAs is currently known to consist of about 20 structurally related water-soluble compounds. MAAs have the basic structure of an amino acid or an amino alcohol, which is combined to a cyclohexenone or a cyclohexenimine. MAAs are considered to have a UV-absorbing function with maximum absorption ranging from 310 to 360 nm, since their levels in some algae depend on the depths of their habitation, and moreover, UV irradiation induces the production of MAAs in the fronds. On the other hand, the levels of MAAs are also sensitive to spawning seasons in some invertebrates, suggesting that they have latent specific functions other than UV-absorption (Bandaranayake et al., 1997). A possible role of MAAs as osmolytes has also been suggested, since MAAs show high solubility in water and have high contents *in vivo* (Shick and Dunlap, 2002; Oren, 1997). *Porphyra* spp. contains a considerable amount of porphyra-334 (λmax = 344 nm) as the major MAA (Takano et al, 1981) (Figure 2). It has been reported that the content of porphyra-334 (from 0.5 to 1.5 g/100 g dry weight) is positively correlated with the quality of the nori product (Ishihara et al., 2008).

Porphyra-334 can be considered a potential natural source for UVA (320-400 nm) protective sunscreens (Torres et al., 2006). Photoprotection is now one of the main goals in social education, in order to reduce the risk of sun exposure and cellular damage. Sunscreens have been used to protect against the acute effects of UV radiation, and they can also have protective effects on chronic UV-induced changes, such as photoaging and skin cancer. The topical application of porphyra-334 (plus shinorine) protected against UV-induced skin damage in mice and contributed to maintaining the antioxidant defense system of the skin as well as the expression of heat shock protein-70 (HSP70) (de la Coba et al., 2009).

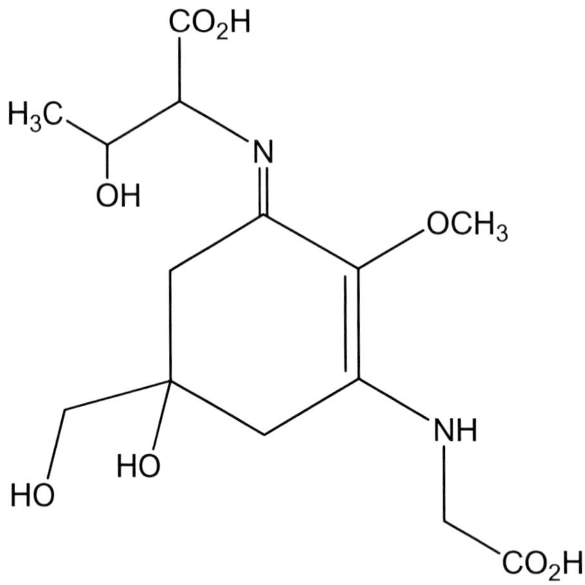

Figure 2. Structure of porphyra-334.

Oyamada et al. demonstrated the protective effect of MAAs, including porphyra-334, on human fibroblasts against UV irradiation (Oyamada et al., 2008). Interestingly, they also found that MAAs promoted the proliferation of human skin fibroblasts. These results indicate that porphyra-334 might have various potential uses in cosmetics and toiletries as a sunscreen and might help stimulate skin cell renewal.

Several authors have reported that MAAs have antioxidative activities. Suh et al. (2003) showed that mycosporine-glycine can effectively quench singlet oxygen. Dunlap and Yamamoto (1995) reported that mycosporine-glycine have moderate antioxidative activities compared with two other water-soluble antioxidants, ascorbic acid and urea. We demonstrated that porphyra-334 has antioxidative activities under various oxidative stresses (Tao et al., 2008). Our results obtained from reaction systems evaluating anti-oxidative activities against linoleic acid showed that porphyra-334 is able to suppress the oxidation induced by alkyl radicals and singlet oxygen. In addition, we found a synergistic effect of α-tocopherol and porphyra-334, but the mechanism of the synergistic activities remains unknown. The storage stability or health effects of nori are at least partly due to these properties of porphyra-334 as an antioxidant and a synergist. Interestingly, Yoshiki et al. (2009) identified a dehydrated compound of porphyra-334 from a water-soluble fraction of *P. yezoensis* after heat treatment over 100°C. This compound, termed NAC (Nori antioxidant compound), exhibited an extremely strong DPPH-radical scavenging activity.

There is no information about the metabolic fate of dietary MAAs in humans. In medaka, the MAAs palythine and asterina-330 were present in significantly greater amounts in the eyes of fish fed an MAA-rich diet compared with control fish (Mason et al., 1998). However, in hairless mice, no dietary MAAs were found in any tissues. It is important to clarify the dietary fate of porphyra-334 and NAC to understand their nutritional and food functions.

PHYCOERYTHRIN

Phycobiliproteins are a family of light-harvesting macromolecules that function as components of the photosynthetic apparatus in cyanobacteria and in several groups of eukaryotic algae, including red algae (Rüdiger, 1994). Phycobiliproteins in phycobilisomes can transfer light energy to chlorophyll a in the type II photosystem at a reaction center in the photosynthetic membrane. Phycoerythrin (λmax ~540-570 nm), which is a brilliantly colored, highly absorbent and fluorescent component, is the major phycobiliprotein in various types of red algae, including *P. yezoensis*. Phycoerythrin contains phycobilin chromophores bound to specific cysteines in the polypeptide chains by thioether bonds. The most representative phycobilin in phycoerythrin is phycoerythrobilin (Figure 3). The main application of phycoerythrin at this time is as a fluorescent marker of cells and macromolecules in biomedical research, and it is used as a highly sensitive fluorescence technique (Telford et al., 2001). In addition, phycoerythrin can be used as a natural dye in foods and cosmetics (Arad and Yaron, 1992).

It has been reported that phycobiliproteins, especially phycocyanin from blue-green algae, have antioxidant activities and also have preventive effects on inflammation and some diseases associated with oxidative stress (Hirata et al., 2000; Romay et al., 2003; Nemoto-Kawamura et al., 2004; Faroop et al., 2004). Phycocyanin can scavenge alkoxyl, hydroxyl and peroxyl radicals and can react with peroxinitrite ($ONOO^-$) and hypochlorous acid (HOCl). In experimental models of inflammation, phycocyanin exerts anti-inflammatory effects. Phycocyanin reduces edema, histamine release, myeloperoxidase activity and levels of prostaglandins and leukotrienes. These anti-inflammatory effects of phycocyanin might be due to its scavenging properties toward reactive oxygen species and its inhibitory effects on cyclooxygenase-2 (COX-2) activity. However, there is a little information about food functions of phycoerythrin from red algae. Soni et al. evaluated the protective action of orally administered phycoerythrin on carbon tetrachloride (CCl_4)-induced hepatic and renal toxicity in rats (Soni et al., 2008). Their results demonstrated that phycoerythrin from red algae ameliorates CCl_4 toxicity caused by reactive oxygen species.

We have evaluated the anti-inflammatory effect of phycoerythrin prepared from *P. yezoensis* (unpublished data). Mast cells play important roles in type I allergic reactions. In mast cells, the cross-linking of IgE-bound FcεRI by antigen initiates intracellular biochemical cascades followed by degranulation (Metcalfe et al., 1997).

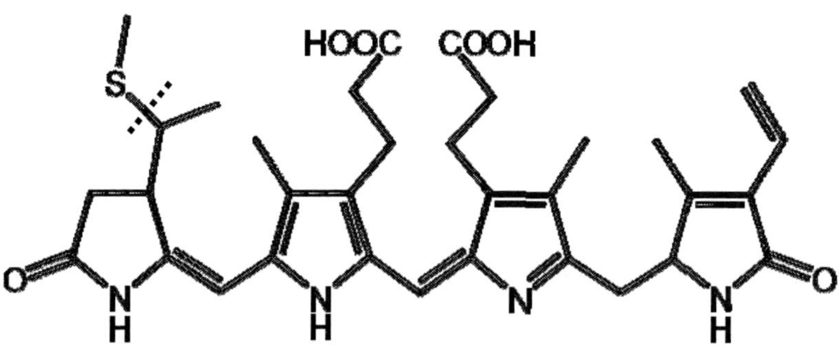

Figure 3. Structure of phycoerythrobilin.

Degranulation results in the release of histamine and many kinds of inflammatory mediators from mast cells which cause allergic reactions. Therefore, the inhibition of degranulation by functional foods and medicinal compounds appears to be a requirement for an anti-allergenic response. β-Hexosaminidase is stored in secretory granules of mast cells and is released concurrently with histamine by the degranulation (Nakano et al., 2005). Thus, β-hexosaminidase activity in the medium is used as a marker of mast cell degranulation. The rat basophilic leukemia cell line (RBL-2H3), a tumor analog and a model for mast cells, has the phenotypic characteristics of mucosal mast cells and is a good tool for studying effects on the activation of mast cells. Phycoerythrin and phycoerythrobilin (which is a chromophore of phycoerythrin) inhibited the calcium ionophore-induced release of β-hexosaminidase by RBL-2H3 cells. However, phycoerythrobilin, but not phycoerythrin, inhibited the antigen-induced degranulation of RBL-2H3 cells. In addition, we found a tendency to reduce the degranulation of peritoneal exudate cells after ingestion of phycoerythrin or phycoerythrobilin in rats. Peritoneal exudate cells have been widely used for studying allergic mediator release, and those cells are easily separated, and contain peritoneal mast cells and basophilic granulocytes (Schmutzler et al., 1995). Our results suggest that dietary phycoerythrin might exhibit an anti-inflammatory activity due to phycoerythrobilin, which has a similar structure to bilirubin after intestinal digestion and absorption. However, there is no information about the fate of dietary phycoerythrin.

Photosensitized oxidation is caused by exposure to sunlight during the traditional process of preparing dried nori, because the fatty acid composition of *P. yezoensis* is rich in PUFAs, mainly eicosapentaenoic acid (EPA), which are labile to oxidation. There are two categories of photosensitized oxidation, Type I and Type II. In Type II, singlet oxygen ($^1O_2$) is generated when excitation is transferred from the sensitizer and this active oxygen molecule reacts with olefinic double bonds by a concerted ene mechanism. Therefore, the isomeric compositions of hydroperoxides formed by the oxidation of PUFAs with $^1O_2$ are different from those generated by autoxidation (Frankel, 1984). Lipid oxidation products, including volatile decompositions generated from specific positional isomers of hydroperoxides by photo-oxidation, might contribute to the quality and flavor of traditional sun-dried nori. We investigated the photosensitized oxidation of EPA in *P. yezoensis* during exposure to light and found that phycoerythrin, at least in part, contributes to the generation of singlet oxygen and the photo-oxidation (Sugawara et al., 2006). The content of phycoerythrin is important for the quality of color, flavor, nutrition and biological functions in *P. yezoensis*.

OTHER COMPOUNDS

Kazlowska et al. (2010) identified catechol, rutin and hesperidin from *P. dentata* as compounds having anti-inflammatory activities. The crude extract and those phenolic compounds inhibited the production of NO in LPS-stimulated RAW 264.7 cells. Catechol is a better suppressor of the up-regulation of the iNOS promoter and NF-□B enhancer than is rutin.

Glycerol galactoside is known to be a component of *Porphyra,* and its content is closely and negatively correlated with the contents of other components (Ishihara et al., 2008). Low-quality nori contains large amounts of glycerol galactoside, more than 10% by dry weight. It has been reported that glycerol galactoside is not digested by any digestive enzyme and is, therefore, not absorbed in the small intestine (Sugawara and Miyazawa, 2000). Muraoka et al. (2008) found that glycerol galactoside from *P. yezoensis* is a substrate for fermentation by bifidobacteria. Their results suggest that glycerol galactoside from low-quality nori can be used as a prebiotic.

Hwang et al. (2008) reported that a 14 kDa protein extracted from *P. yezoensis* has chemoprotective effects on acetaminophen-induced liver injury in rats. When added to the diet, this protein inhibited acetaminophen-induced the depletion of GSH and the activation of caspase-3 in the liver.

Potent inhibitors of angiotensin I-converting enzyme (ACE) have been identified in *P. yezoensis* (Suetsuna, 1998). ACE participates in regulating blood pressure in the renin-angiotensin system, and inhibitors of ACE have been used as antihypertensive drugs. Recently, many ACE inhibitor peptides have been isolated from various foods, and Suetsuna found that a pepsin digest of *P. yezoensis* has potent ACE inhibitory activity.

Conclusion

In this chapter, we reviewed the nutritional and physiological functions of components that are distinctive of *Porphyra* spp.. It is important to understand the nutritional roles and functions of *Porphyra*, because algae are a natural source of dietary compounds that differ from those of terrestrial animals and plants.

References

Arad S; Yaron A. Natural pigments from red microalgae for use in foods and cosmetics. *Trends Food Sci. Technol.* 1992, 3, 92-97

Bandaranayake WM; Bourne DJ; Sim RG. Chemical composition during maturing and spawning of the sponge *Dysidea herbacea* (Porifera: Demospongiae). *Comp. Biochem. Physiol.* 1997, 118B, 851–859.

Bocanegra A; Nieto A; Bastida S; Benedí J; Sánchez-Muniz FJ. A Nori but not a Konbu, dietary supplement decreases the cholesterolaemia, liver fat infiltration and mineral bioavailability in hypercholesterolaemic growing Wistar rats. *Br. J. Nutr.* 2008, 99, 272-280.

Dagnelie PC; van Staveren WA; van den Berg H. Vitamin B-12 from algae appears not to be bio-available. *Am. J. Clin. Nutr.* 1991, 53, 695-697.

Dawczynski C; Schäfer U; Leiterer M; Jahreis G. Nutritional and toxicological importance of macro, trace, and ultra-trace elements in algae food products. *J. Agric. Food Chem.* 2007, 55, 10470-10475.

de la Coba F; Aguilera J; de Gálvez MV; Alvarez M; Gallego E; Figueroa FL; Herrera E. Prevention of the ultraviolet effects on clinical and histopathological changes, as well as the heat shock protein-70 expression in mouse skin by topical application of algal UV-absorbing compounds. *J. Dermatol. Sci.* 2009, 55, 161-169.

Dunlap WC; Yamamoto Y. Small-molecule antioxidant in marine organisms: antioxidant activity of mycosporine-glycine. *Comp. Biochem. Physiol.* 1995, 112, 105-114.

Farooq SM; Asokan D; Sakthivel R; Kalaiselvi P; Varalakshmi P. Salubrious effect of C-phycocyanin against oxalate-mediated renal cell injury. *Clin. Chim. Acta.* 2004, 348, 199-205.

Frankel EN. Chemistry of free radical and singlet oxidation of lipids. *Prog. Lipid Res.* 1984, 23, 197-221.

García-Casal MN; Pereira AC; Leets I; Ramírez J; Quiroga MF. High iron content and bioavailability in humans from four species of marine algae. *J. Nutr.* 2007, 137, 2691-2695.

Goñi I; Gudiel-Urbano M; Bravo L; Saura-Calixto F. Dietary modulation of bacterial fermentative capacity by edible seaweeds in rats. *J. Agric. Food Chem.* 2001, 49, 2663-2668.

Gudiel-Urbano M; Goñi I. Bioavailability of nutrients in rats fed on edible seaweeds, Nori (Porphyra tenera) and Wakame (Undaria pinnatifida), as a source of dietary fiber. *Food Chem.* 2002a, 76, 281-286.

Gudiel-Urbano M; Goñi I. Effect of edible seaweeds (Undaria pinnatifida and Porphyra ternera) on the metabolic activities of intestinal microflora in rats. *Nutr. Res.* 2002b, 22, 323-331.

Guo TT; Xu HL; Zhang LX; Zhang JP; Guo YF; Gu JW; He PM. *In vivo* protective effect of *Porphyra yezoensis* polysaccharide against carbon tetrachloride induced hepatotoxicity in mice. *Regul. Toxicol. Pharmacol.* 2007, 49, 101-106.

Hehemann JH; Correc G; Barbeyron T; Helbert W; Czjzek M; Michel G. Transfer of carbohydrate-active enzymes from marine bacteria to Japanese gut microbiota. *Nature.* 2010, 464, 908-912.

Hirata T; Tanaka M; Ooike M; Tsunomura T; Sakaguchi M. Antioxidant activities of phycocyanobilin prepared from Spirulina platensis. *J. Appl. Phycol.* 2000, 12, 435-439.

Hwang HJ; Kwon MJ; Kim IH; Nam TJ. Chemoprotective effects of a protein from the red algae *Porphyra yezoensis* on acetaminophen-induced liver injury in rats. *Phytother. Res.* 2008, 22, 1149-1153.

Inoue N; Yamano N; Sakata K; Nagao K; Hama Y; Yanagita T. The sulfated polysaccharide porphyran reduces apolipoprotein B100 secretion and lipid synthesis in HepG2 cells. *Biosci. Biotechnol. Biochem.* 2009, 73, 447-449.

Ishihara K; Oyamada C; Matsushima R; Murata M; Muraoka T. Inhibitory effect of porphyran, prepared from dried "Nori", on contact hypersensitivity in mice. *Biosci. Biotechnol. Biochem.* 2005, 69, 1824-1830.

Ishihara K; Oyamada C; Sato Y; Danno H; Kimiya T; Kaneniwa M; Kunitake H; Muraoka T. Relationships between quality parameters and content of glycerol galactoside and porphyra-334 in dried nori *Porphyra yezoensis*. *Fish. Sci.* 2008, 74, 167-173.

Kazłowska K; Hsu T; Hou CC; Yang WC; Tsai GJ. Anti-inflammatory properties of phenolic compounds and crude extract from *Porphyra dentata. J. Ethnopharmacol.* 2010, 128, 123-130.

Kwon MJ; Nam TJ. Porphyran induces apoptosis related signal pathway in AGS gastric cancer cell lines. *Life Sci.* 2006, 79, 1956-1962.

Mason DS; Schafer F; Shick JM; Dunlap WC. Ultraviolet radiation-absorbing mycosporine-like amino acids (MAAs) are acquired from their diet by medaka fish (*Oryzias latipes*) but not by SKH-1 hairless mice. *Comp. Biochem. Physiol. A.* 1998, 120, 587-598.

Metcalfe DD; Baram D; Mekori YA. Mast cells. *Physiol. Rev.* 1997, 77, 1033-1079.

Miyamoto E; Yabuta Y; Kwak CS; Enomoto T; Watanabe F. Characterization of vitamin B_{12} compounds from Korean purple laver (*Porphyra* sp.) products. *J. Agric. Food Chem.* 2009, 57, 2793-2796.

Morita K; Tobiishi K. Increasing effect of nori on the fecal excretion of dioxin by rats. *Biosci. Biotechnol. Biochem.* 2002, 66, 2306-2313.

Morrice LM; McLean MW; Long W.; Williamson FB. Porphyran primary structure, *Eur. J. Biochem.*, 1983, 133, 673-684.

Muraoka T; Ishihara K; Oyamada C; Kunitake H; Hirayama I; Kimura T. Fermentation properties of low-quality red alga susabinori *Porphyra yezoensis* by intestinal bacteria. *Biosci. Biotechnol. Biochem.* 2008, 72, 1731-1739.

Nakano N; Nakao A; Uchida T; Shirasaka N; Yoshizumi H; Okumura K; Tsuboi R; Ogawa H. Effects of arakidonic acid analogs on Fc□RI-mediated activation of mast cells. *Biochim. Biophys. Acta* 2005, 1738, 19-28.

Nemoto-Kawamura C; Hirahashi T; Nagai T; Yamada H; Katoh T; Hayashi O. Phycocyanin enhances secretary IgA antibody response and suppresses allergic IgE antibody response in mice immunized with antigen-entrapped biodegradable microparticles. *J. Nutr. Sci. Vitaminol.* 2004, 50, 129-136.

Noda H; Amano H; Arashima K; Nisizawa K. Antitumor activity of marine algae. *Hydrobiologia.* 1990, 204/205, 577-584.

Okai Y; Higashi-Okai K; Nakamura S; Yano Y; Otani S. Suppressive effects of the extracts of Japanese edible seaweeds on mutagen-induced umu C gene expression in Salmonella typhimurium (TA 1535/pSK 1002) and tumor promotor-dependent ornithine decarboxylase induction in BALB/c 3T3 fibroblast cells. *Cancer Lett.* 1994, 87, 25-32.

Oren A. Mycosporine-like amino acids as osmotic solutes in a community of halophilic cyanobacteria. *Biomicrobiol. J.* 1997, 14, 231-240.

Oyamada C; Kaneniwa M; Ebitani K; Murata M; Ishihara K. Mycosporine-like amino acids extracted from scallop (*Patinopecten yessoensis*) ovaries: UV protection and growth stimulation activities on human cells. *Mar. Biotechnol. (NY).* 2008, 10, 141-150.

Rauma AL; Törrönen R; Hänninen O; Mykkänen H. Vitamin B-12 status of long-term adherents of a strict uncooked vegan diet ("living food diet") is compromised. *J. Nutr.* 1995, 125, 2511-2515.

Romay C; González R; Ledón N; Remirez D; Rimbau V. C-phycocyanin: a biliprotein with antioxidant, anti-inflammatory and neuroprotective effects. *Curr. Protein Pept. Sci.* 2003, 4, 207-216.

Rüdiger W. Phycobiliproteins and phycobilins. *Prog. Phycol. Res.* 1994, 10, 97-135.

Rupérez P. Mineral content of edible marine seaweeds. *Food Chem.* 2002, 79, 23-26.

Schmutzler W; Bolsmann K; Zwadlo-Klarwasser G Comparison of histamine release from human blood monocytes, lymphocytes, adenosidal and skin mast cells. *Int. Arch. Allergy Immunol.* 1995, 107, 194-196.

Shick JM; Dunlap WC. Mycosporine-like amino acids and related gadusols: biosynthesis, accumulation, and UV-protective functions in aquatic organisms. *Annu. Rev. Physiol.* 2002, 64, 223-262.

Soni B; Visavadiya NP; Madamwar D. Ameliorative action of cyanobacterial phycoerythrin on CCl_4-induced toxicity in rats. *Toxicology.* 2008, 248, 59-65.

Suetsuna K. Purification and identification of angiotensin I-converting enzyme inhibitors from the red alga *Porphyra yezoensis*. *J. Mar. Biotechnol.* 1998, 6, 163-167.

Sugawara T; Komura Y; Hagino H; Hirata T. Phycoerythrin contributes to the photo-oxidation of eicosapentaenoic acid in *Porphyra yezoenisis* during light-exposure, *J. Food Sci.* 2006, 71, S486-S491.

Sugawara T; Miyazawa T. Digestion of plant monogalactosyldiacylglycerol and digalactosyldiacylglycerol in rat alimentary canal. *J. Nutr. Biochem.* 2000, 11, 147-152

Suh HJ; Lee HW; Jung J. Mycosporine-glycine protects biological systems against photodynamic damage by quenching singlet oxygen with a high efficiency. *Photochem. Photobiol.* 2003, 78, 109-113.

Takano S; Nakanishi A; Uemura D; Hirata T. Isolation and structure of a 334 nm UV-absorbing substance, porphyra-334 from the alga *Porphyra tenera* Kjellman. *Chem. Lett.* 1981, 10, 1413-1414.

Takenaka S; Sugiyama S; Ebara S; Miyamoto E; Abe K; Tamura Y; Watanabe F; Tsuyama S; Nakano Y. Feeding dried purple laver (nori) to vitamin B_{12}-deficient rats significantly improves vitamin B_{12} status. *Br. J. Nutr.* 2001, 85, 699-703.

Tao C; Sugawara T; Maeda S; Wang X; Hirata T. Antioxidative activities of a mycosporine-like amino acid, porphyra-334. *Fish. Sci.* 2008, 74, 1166-1172.

Telford WG; Moss MW; Morseman JP; Allnutt FC. Cryptomonad algal phycobiliproteins as fluorochromes for extracellular and intracellular antigen detection by flow cytometry. *Cytometry.* 2001, 44, 16-23.

Torres A; Enk CD; Hochberg M; Srebnik M. Porphyra-334, a potential natural source for UVA protective sunscreens. *Photochem. Photobiol. Sci.* 2006, 5, 432-435.

Tsuge K; Okabe M; Yoshimura T; Sumi T; Tachibana H; Yamada K. Dietary effects of porphyran from *Porphyra yezoensis* on growth and lipid metabolism of Sprague-Dawley rats. *Food Sci. Technol. Res.* 2004, 10, 147-151.

van den Berg H; Dagnelie PC; van Staveren WA. Vitamin B12 and seaweed. *Lancet,* 1988, 1, 242-243.

Watanabe F; Takenaka S; Katsura H; Masumder SA; Abe K; Tamura Y; Nakano Y. Dried green and purple lavers (Nori) contain substantial amounts of biologically active vitamin B(12) but less of dietary iodine relative to other edible seaweeds. *J. Agric. Food Chem.* 1999, 47, 2341-2343.

Yamamoto I; Maruyama H. Effect of dietary seaweed preparations on 1,2-dimethylhydrazine-induced intestinal carcinogenesis in rats. *Cancer Lett.* 1985, 26, 241-251.

Yamamoto I; Maruyama H; Moriguchi M. The effect of dietary seaweeds on 7,12-dimethyl-benz[a]anthracene-induced mammary tumorigenesis in rats. *Cancer Lett.* 1987, 35, 109-118.

Yang YJ; Nam SJ; Kong G; Kim MK. A case-control study on seaweed consumption and the risk of breast cancer. *Br. J. Nutr.* 2010, 103, 1345-1353.

Yoshiki M; Tsuge K; Tsuruta Y; Yoshimura T.; Koganemura K; Sumi T; Matsui T; Matsumoto K. Production of new antioxidant compound from mycosporine-like amino acid, porphyra-334 by heat treatment. *Food Chem.* 2009, 1127-1132.

Yoshizawa Y; Ametani A; Tsunehiro J; Nomura K; Itoh M; Fukui F; Kaminogawa S. Macrophage stimulation activity of the polysaccharide fraction from a marine alga

(*Porphyra yezoensis*): structure-function relationships and improved solubility. *Biosci. Biotechnol. Biochem.* 1995, 59, 1933-1937.

Yoshizawa Y; Enomoto A; Todoh H; Ametani A; Kaminogawa S. Activation of murine macrophages by polysaccharide fractions from marine algae (*Porphyra yezoensis*). *Biosci. Biotechnol. Biochem.* 1993, 57, 1862-1866.

Yuan YV; Walsh NA. Antioxidant and antiproliferative activities of extracts from a variety of edible seaweeds. *Food Chem. Toxicol.* 2006, 44, 1144-1150.

Zhang Q; Yu P; Li Z; Zhang H;, Xu Z; Li P. Antioxidant activities of sulfated polysaccharide fractions from *Porphyra haitanesis. J. Appl. Phycol.* 2003a, 15, 305-310.

Zhang Q; Li N; Zhou G; Lu X; Xu Z; Li Z. In vivo antioxidant activity of polysaccharide fraction from *Porphyra haitanesis* (Rhodephyta) in aging mice. *Pharmacol. Res.* 2003b, 48, 151–155.

Zhang Q; Li N; Liu X; Zhao Z; Li Z; Xu Z. The structure of a sulfated galactan from *Porphyra haitanensis* and its *in vivo* antioxidant activity. *Carbohydr. Res.* 2004, 339, 105-111.

Zhang Z; Zhang Q; Wang J.; Zhang H; Niu X; Li P. Preparation of the different derivatives of the low-molecular-weight porphyran from *Porphyra haitanensis* and their antioxidant activities *in vitro*. *Int. J. Biol. Macromolecul.* 2009, 45, 22-26.

Zhao T; Zhang Q; Qi H; Zhang H; Niu X; Xu Z; Li Z. Degradation of porphyran from *Porphyra haitanensis* and the antioxidant activities of the degraded porphyrans with different molecular weight. *Int. J. Biol. Macromol.* 2006, 38, 45-50.

INDEX

A

B

C

D

E

F

G

H

I

N

O

P

Q

R

S

T

U

V

W

X

Y

Z